OXFORD LOGIC GUIDES

Series Editors

A.J. MACINTYRE   D.S. SCOTT

Emeritus Editors

D.M. Gabbay
John Shepherdson

T0177735

OXFORD LOGIC GUIDES

For a full list of titles please visit
http://www.oup.co.uk/academic/science/maths/series/OLG/

Category Theory

Second Edition

STEVE AWODEY
Carnegie Mellon University

OXFORD
UNIVERSITY PRESS

Great Clarendon Street, Oxford OX2 6DP

Oxford University Press is a department of the University of Oxford.
It furthers the University's objective of excellence in research, scholarship,
and education by publishing worldwide in

Oxford New York

Auckland   Cape Town Dar es Salaam   Hong Kong   Karachi
Kuala Lumpur Madrid Melbourne Mexico City   Nairobi
New Delhi Shanghai Taipei Toronto
With offices in
Argentina Austria Brazil Chile Czech Republic France Greece
Guatemala Hungary Italy Japan South Korea Poland Portugal
Singapore Switzerland Thailand Turkey Ukraine Vietnam

Oxford is a registered trade mark of Oxford University Press
in the UK and in certain other countries

Published in the United States
by Oxford University Press Inc., New York

ISBN 978-0-19-923718-0

Printed and bound by CPI Group (UK) Ltd, Croydon, CRO 4YY

in memoriam
Saunders Mac Lane

PREFACE TO THE SECOND EDITION

This second edition of *Category Theory* differs from the first in two respects: firstly, numerous corrections and revisions have been made to the text, including correcting typographical errors, revising details in exposition and proofs, providing additional diagrams, and finally adding an entirely new section on monoidal categories. Secondly, dozens of new exercises were added to make the book more useful as a course text and for self-study. To the same end, solutions to selected exercises have also been provided; for these, I am grateful to Spencer Breiner and Jason Reed.

Steve Awodey
Pittsburgh
September 2009

PREFACE

Why write a new textbook on Category Theory, when we already have Mac Lane's *Categories for the Working Mathematician*? Simply put, because Mac Lane's book is for the working (and aspiring) mathematician. What is needed now, after 30 years of spreading into various other disciplines and places in the curriculum, is a book for everyone else.

This book has grown from my courses on Category Theory at Carnegie Mellon University over the last 10 years. In that time, I have given numerous lecture courses and advanced seminars to undergraduate and graduate students in Computer Science, Mathematics, and Logic. The lecture course based on the material in this book consists of two, 90-minute lectures a week for 15 weeks. The germ of these lectures was my own graduate student notes from a course on Category Theory given by Mac Lane at the University of Chicago. In teaching my own course, I soon discovered that the mixed group of students at Carnegie Mellon had very different needs than the Mathematics graduate students at Chicago and my search for a suitable textbook to meet these needs revealed a serious gap in the literature. My lecture notes evolved over a time to fill this gap, supplementing and eventually replacing the various texts I tried using.

The students in my courses often have little background in Mathematics beyond a course in Discrete Math and some Calculus or Linear Algebra or a course or two in Logic. Nonetheless, eventually, as researchers in Computer Science or Logic, many will need to be familiar with the basic notions of Category Theory, without the benefit of much further mathematical training. The Mathematics undergraduates are in a similar boat: mathematically talented, motivated to learn the subject by its evident relevance to their further studies, yet unable to follow Mac Lane because they still lack the mathematical prerequisites. Most of my students do not know what a free group is (yet), and so they are not illuminated to learn that it is an example of an adjoint.

This, then, is intended as a text and reference book on Category Theory, not only for students of Mathematics, but also for researchers and students in Computer Science, Logic, Linguistics, Cognitive Science, Philosophy, and any of the other fields that now make use of it. The challenge for me was to make the basic definitions, theorems, and proof techniques understandable to this readership, and thus without presuming familiarity with the main (or at least original) applications in algebra and topology. It will not do, however, to develop the subject in a vacuum, simply skipping the examples and applications. Material at this level of abstraction is simply incomprehensible without the applications and examples that bring it to life.

Faced with this dilemma, I have adopted the strategy of developing a few basic examples from scratch and in detail—namely posets and monoids—and then carrying them along and using them throughout the book. This has several didactic advantages worth mentioning: both posets and monoids are themselves special kinds of categories, which in a certain sense represent the two "dimensions" (objects and arrows) that a general category has. Many phenomena occurring in categories can best be understood as generalizations from posets or monoids. On the other hand, the categories of posets (and monotone maps) and monoids (and homomorphisms) provide two further, quite different examples of categories in which to consider various concepts. The notion of a limit, for instance, can be considered both in a given poset and in the category of posets.

Of course, many other examples besides posets and monoids are treated as well. For example, the chapter on groups and categories develops the first steps of Group Theory up to kernels, quotient groups, and the homomorphism theorem, as an example of equalizers and coequalizers. Here, and occasionally elsewhere (e.g., in connection with Stone duality), I have included a bit more Mathematics than is strictly necessary to illustrate the concepts at hand. My thinking is that this may be the closest some students will ever get to a higher Mathematics course, so they should benefit from the labor of learning Category Theory by reaping some of the nearby fruits.

Although the mathematical prerequisites are substantially lighter than for Mac Lane, the standard of rigor has (I hope) not been compromised. Full proofs of all important propositions and theorems are given, and only occasional routine lemmas are left as exercises (and these are then usually listed as such at the end of the chapter). The selection of material was easy. There is a standard core that must be included: categories, functors, natural transformations, equivalence, limits and colimits, functor categories, representables, Yoneda's lemma, adjoints, and monads. That nearly fills a course. The only "optional" topic included here is cartesian closed categories and the λ-calculus, which is a must for computer scientists, logicians, and linguists. Several other obvious further topics were purposely not included: 2-categories, topoi (in any depth), and monoidal categories. These topics are treated in Mac Lane, which the student should be able to read after having completed the course.

Finally, I take this opportunity to thank Wilfried Sieg for his exceptional support of this project; Peter Johnstone and Dana Scott for helpful suggestions and support; André Carus for advice and encouragement; Bill Lawvere for many very useful comments on the text; and the many students in my courses who have suggested improvements to the text, clarified the content with their questions, tested all of the exercises, and caught countless errors and typos. For the latter, I also thank the many readers who took the trouble to collect and send helpful corrections, particularly Brighten Godfrey, Peter Gumm, Bob Lubarsky, and Dave Perkinson. Andrej Bauer and Kohei Kishida are to be thanked for providing Figures 9.1 and 8.1, respectively. Of course, Paul Taylor's macros for commutative diagrams must also be acknowledged. And my dear Karin deserves thanks for too many things to mention. Finally, I wish to record here my debt of gratitude to my mentor Saunders Mac Lane, not only for teaching me Category Theory, and trying to teach me how to write, but also for helping me to find my place in Mathematics. I dedicate this book to his memory.

Steve Awodey
Pittsburgh
September 2005

CONTENTS

1

CATEGORIES

1.1 Introduction

What is category theory? As a first approximation, one could say that category theory is the mathematical study of (abstract) *algebras of functions*. Just as group theory is the abstraction of the idea of a system of permutations of a set or symmetries of a geometric object, category theory arises from the idea of a system of functions among some objects.

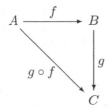

We think of the composition $g \circ f$ as a sort of "product" of the functions f and g, and consider abstract "algebras" of the sort arising from collections of functions. A category is just such an "algebra," consisting of objects $A, B, C, \ldots$ and arrows $f : A \to B$, $g : B \to C, \ldots$, that are closed under composition and satisfy certain conditions typical of the composition of functions. A precise definition is given later in this chapter.

A branch of abstract algebra, category theory was invented in the tradition of Felix Klein's *Erlanger Programm*, as a way of studying and characterizing different kinds of mathematical structures in terms of their "admissible trans-formations." The general notion of a category provides a characterization of the notion of a "structure-preserving transformation," and thereby of a species of structures admitting such transformations.

The historical development of the subject has been, very roughly, as follows:

1945 Eilenberg and Mac Lane's "General theory of natural equivalences" was the original paper, in which the theory was first formulated.

Late 1940s The main applications were originally in the fields of algebraic topology, particularly homology theory, and abstract algebra.

1950s A. Grothendieck et al. began using category theory with great success in algebraic geometry.

1960s F.W. Lawvere and others began applying categories to logic, revealing some deep and surprising connections.

1970s Applications were already appearing in computer science, linguistics, cognitive science, philosophy, and many other areas.

One very striking thing about the field is that it has such wide-ranging applications. In fact, it turns out to be a kind of universal mathematical language like set theory. As a result of these various applications, category theory also tends to reveal certain connections between different fields—like logic and geometry. For example, the important notion of an *adjoint functor* occurs in logic as the existential quantifier and in topology as the image operation along a continuous function. From a categorical point of view, these turn out to be essentially the same operation.

The concept of adjoint functor is in fact one of the main things that the reader should take away from the study of this book. It is a strictly category-theoretical notion that has turned out to be a conceptual tool of the first magnitude—on par with the idea of a continuous function.

In fact, just as the idea of a topological space arose in connection with continuous functions, so also the notion of a category arose in order to define that of a functor, at least according to one of the inventors. The notion of a functor arose—so the story goes on—in order to define natural transformations. One might as well continue that natural transformations serve to define adjoints:

Category

Functor

Natural transformation

Adjunction

Indeed, that gives a good outline of this book.

Before getting down to business, let us ask why it should be that category theory has such far-reaching applications. Well, we said that it is the abstract theory of functions, so the answer is simply this:

Functions are everywhere!

And everywhere that functions are, there are categories. Indeed, the subject might better have been called *abstract function theory*, or, perhaps even better: *archery*.

1.2 Functions of sets

We begin by considering functions between sets. I am not going to say here what a function is, anymore than what a set is. Instead, we will assume a working knowledge of these terms. They can in fact be *defined* using category theory, but that is not our purpose here.

Let f be a function from a set A to another set B, we write

$$f : A \to B.$$

To be explicit, this means that f is defined on all of A and all the values of f are in B. In set theoretic terms,

$$\text{range}(f) \subseteq B.$$

Now suppose we also have a function $g : B \to C$,

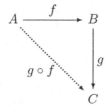

then there is a composite function $g \circ f : A \to C$, given by

$$(g \circ f)(a) = g(f(a)) \qquad a \in A. \tag{1.1}$$

Now this operation "$\circ$" of composition of functions is associative, as follows. If we have a further function $h : C \to D$

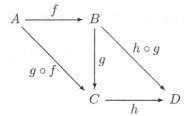

and form $h \circ g$ and $g \circ f$, then we can compare $(h \circ g) \circ f$ and $h \circ (g \circ f)$ as indicated in the diagram given above. It turns out that these two functions are always identical,

$$(h \circ g) \circ f = h \circ (g \circ f)$$

since for any $a \in A$, we have

$$((h \circ g) \circ f)(a) = h(g(f(a))) = (h \circ (g \circ f))(a)$$

using (1.1).

By the way, this is, of course, what it means for two functions to be equal: for every argument, they have the same value.

Finally, note that every set A has an identity function

$$1_A : A \to A$$

given by

$$1_A(a) = a.$$

These identity functions act as "units" for the operation $\circ$ of composition, in the sense of abstract algebra. That is to say,

$$f \circ 1_A = f = 1_B \circ f$$

for any $f : A \to B$.

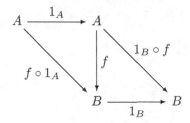

These are all the properties of set functions that we want to consider for the *abstract* notion of function: composition and identities. Thus, we now want to "abstract away" everything else, so to speak. That is what is accomplished by the following definition.

1.3 Definition of a category

Definition 1.1. A *category* consists of the following data:

- *Objects*: $A, B, C, \ldots$
- *Arrows*: $f, g, h, \ldots$
- For each arrow f, there are given objects

$$\mathrm{dom}(f), \qquad \mathrm{cod}(f)$$

called the *domain* and *codomain* of f. We write

$$f : A \to B$$

to indicate that $A = \mathrm{dom}(f)$ and $B = \mathrm{cod}(f)$.

- Given arrows $f : A \to B$ and $g : B \to C$, that is, with

$$\mathrm{cod}(f) = \mathrm{dom}(g)$$

there is given an arrow

$$g \circ f : A \to C$$

called the *composite* of f and g.

- For each object A, there is given an arrow

$$1_A : A \to A$$

called the *identity arrow* of A.

These data are required to satisfy the following laws:

- Associativity:

$$h \circ (g \circ f) = (h \circ g) \circ f$$

for all $f : A \to B$, $g : B \to C$, $h : C \to D$.
- Unit:

$$f \circ 1_A = f = 1_B \circ f$$

for all $f : A \to B$.

A category is *anything* that satisfies this definition—and we will have plenty of examples very soon. For now I want to emphasize that, unlike in Section 1.2, the objects do not have to be sets and the arrows need not be functions. In this sense, a category is an *abstract* algebra of functions, or "arrows" (sometimes also called "morphisms"), with the composition operation "$\circ$" as primitive. If you are familiar with groups, you may think of a category as a sort of generalized group.

1.4 Examples of categories

1. We have already encountered the category **Sets** of sets and functions. There is also the category

$$\textbf{Sets}_{\text{fin}}$$

of all finite sets and functions between them.

Indeed, there are many categories like this, given by restricting the sets that are to be the objects and the functions that are to be the arrows. For example, take finite sets as objects and injective (i.e., "1 to 1") functions as arrows. Since injective functions compose to give an injective function, and since the identity functions are injective, this also gives a category.

What if we take sets as objects and as arrows, those $f : A \to B$ such that for all $b \in B$, the subset

$$f^{-1}(b) \subseteq A$$

has at most two elements (rather than one)? Is this still a category? What if we take the functions such that $f^{-1}(b)$ is finite? infinite? There are lots of such restricted categories of sets and functions.

2. Another kind of example one often sees in mathematics is categories of *structured sets*, that is, sets with some further "structure" and functions that "preserve it," where these notions are determined in some independent way. Examples of this kind you may be familiar with are

- groups and group homomorphisms,
- vector spaces and linear mappings,
- graphs and graph homomorphisms,
- the real numbers $\mathbb{R}$ and continuous functions $\mathbb{R} \to \mathbb{R}$,
- open subsets $U \subseteq \mathbb{R}$ and continuous functions $f : U \to V \subseteq \mathbb{R}$ defined on them,
- topological spaces and continuous mappings,
- differentiable manifolds and smooth mappings,
- the natural numbers $\mathbb{N}$ and all recursive functions $\mathbb{N} \to \mathbb{N}$, or as in the example of continuous functions, one can take partial recursive functions defined on subsets $U \subseteq \mathbb{N}$,
- posets and monotone functions.

Do not worry if some of these examples are unfamiliar to you. Later on, we take a closer look at some of them. For now, let us just consider the last of the above examples in more detail.

3. A partially ordered set or *poset* is a set A equipped with a binary relation $a \leq_A b$ such that the following conditions hold for all $a, b, c \in A$:

   reflexivity: $a \leq_A a$,
   transitivity: if $a \leq_A b$ and $b \leq_A c$, then $a \leq_A c$,
   antisymmetry: if $a \leq_A b$ and $b \leq_A a$, then $a = b$.

For example, the real numbers $\mathbb{R}$ with their usual ordering $x \leq y$ form a poset that is also *linearly* ordered: either $x \leq y$ or $y \leq x$ for any x, y.

An arrow from a poset A to a poset B is a function

$$m : A \to B$$

that is *monotone*, in the sense that, for all $a, a' \in A$,

$$a \leq_A a' \quad \text{implies} \quad m(a) \leq_B m(a').$$

What does it take for this to be a category? We need to know that $1_A : A \to A$ is monotone, but that is clear since $a \leq_A a'$ implies $a \leq_A a'$. We also need to know that if $f : A \to B$ and $g : B \to C$ are monotone, then $g \circ f : A \to C$ is monotone. This also holds, since $a \leq a'$ implies

$f(a) \leq f(a')$ implies $g(f(a)) \leq g(f(a'))$ implies $(g \circ f)(a) \leq (g \circ f)(a')$.
Therefore, we have the category **Pos** of posets and monotone functions.

4. The categories that we have been considering so far are examples of what
   are sometimes called *concrete categories*. Informally, these are categories in
   which the objects are sets, possibly equipped with some structure, and the
   arrows are certain, possibly structure-preserving, functions (we shall see
   later on that this notion is not entirely coherent; see remark 1.7). But in
   fact, one way of understanding what category theory is all about is "doing
   without elements," and replacing them by arrows instead. Let us now take
   a look at some examples where this point of view is not just optional, but
   essential.

   Let **Rel** be the following category: take sets as objects and take binary
   relations as arrows. That is, an arrow $f : A \rightarrow B$ is an arbitrary subset
   $f \subseteq A \times B$. The identity arrow on a set A is the identity relation,

   $$1_A = \{(a, a) \in A \times A \mid a \in A\} \subseteq A \times A.$$

   Given $R \subseteq A \times B$ and $S \subseteq B \times C$, define composition $S \circ R$ by

   $$(a, c) \in S \circ R \quad \text{iff} \quad \exists b.\ (a, b) \in R \ \& \ (b, c) \in S$$

   that is, the "relative product" of S and R. We leave it as an exercise to
   show that **Rel** is in fact a category. (What needs to be done?)

   For another example of a category in which the arrows are not "functions," let the objects be finite sets A, B, C and an arrow $F : A \rightarrow B$ is
   a rectangular matrix $F = (n_{ij})_{i<a, j<b}$ of natural numbers with $a = |A|$
   and $b = |B|$, where $|C|$ is the number of elements in a set C. The com-
   position of arrows is by the usual matrix multiplication, and the identity
   arrows are the usual unit matrices. The objects here are serving simply to
   ensure that the matrix multiplication is defined, but the matrices are not
   functions between them.

5. *Finite categories*
   Of course, the objects of a category do not have to be sets, either. Here
   are some very simple examples:

   - The category **1** looks like this:

     $$*$$

     It has one object and its identity arrow, which we do not draw.

   - The category **2** looks like this:

     $$* \longrightarrow \star$$

     It has two objects, their required identity arrows, and exactly one
     arrow between the objects.

- The category **3** looks like this:

It has three objects, their required identity arrows, exactly one arrow from the first to the second object, exactly one arrow from the second to the third object, and exactly one arrow from the first to the third object (which is therefore the composite of the other two).

- The category **0** looks like this:

It has no objects or arrows.

As above, we omit the identity arrows in drawing categories from now on.

It is easy to specify finite categories—just take some objects and start putting arrows between them, but make sure to put in the necessary identities and composites, as required by the axioms for a category. Also, if there are any loops, then they need to be cut off by equations to keep the category finite. For example, consider the following specification:

$$A \underset{g}{\overset{f}{\rightleftarrows}} B$$

Unless we stipulate an equation like $gf = 1_A$, we will end up with infinitely many arrows gf, $gfgf$, $gfgfgf$, ... This is still a category, of course, but it is not a *finite* category. We come back to this situation when we discuss free categories later in this chapter.

6. One important slogan of category theory is

   It's the arrows that really matter!

Therefore, we should also look at the arrows or "mappings" between categories. A "homomorphism of categories" is called a functor.

Definition 1.2. A *functor*

$$F : \mathbf{C} \to \mathbf{D}$$

between categories **C** and **D** is a mapping of objects to objects and arrows to arrows, in such a way that

(a) $F(f : A \to B) = F(f) : F(A) \to F(B)$,

(b) $F(1_A) = 1_{F(A)}$,

(c) $F(g \circ f) = F(g) \circ F(f)$.

That is, F preserves domains and codomains, identity arrows, and compostion. A functor $F : \mathbf{C} \to \mathbf{D}$ thus gives a sort of "picture"—perhaps distorted—of $\mathbf{C}$ in $\mathbf{D}$.

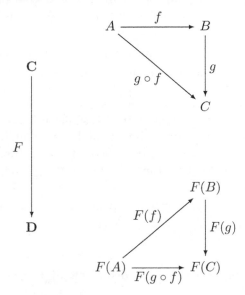

Now, one can easily see that functors compose in the expected way, and that every category $\mathbf{C}$ has an identity functor $1_{\mathbf{C}} : \mathbf{C} \to \mathbf{C}$. So we have another example of a category, namely $\mathbf{Cat}$, the category of all categories and functors.

7. A *preorder* is a set P equipped with a binary relation $p \leq q$ that is both reflexive and transitive: $a \leq a$, and if $a \leq b$ and $b \leq c$, then $a \leq c$. Any preorder P can be regarded as a category by taking the objects to be the elements of P and taking a unique arrow,

$$a \to b \quad \text{if and only if} \quad a \leq b. \tag{1.2}$$

The reflexive and transitive conditions on $\leq$ ensure that this is indeed a category.

Going in the other direction, any category with at most one arrow between any two objects determines a preorder, simply by defining a binary relation $\leq$ on the objects by (1.2).

8. A poset is evidently a preorder satisfying the additional condition of anti-symmetry: if $a \leq b$ and $b \leq a$, then $a = b$. So, in particular, a poset is also a category. Such *poset categories* are very common; for example, for

any set X, the powerset $\mathcal{P}(X)$ is a poset under the usual inclusion relation $U \subseteq V$ between the subsets U, V of X.

What is a functor $F : P \to Q$ between poset categories P and Q? It must satisfy the identity and composition laws.... Clearly, these are just the monotone functions already considered above.

It is often useful to think of a category as a kind of *generalized poset*, one with "more structure" than just $p \leq q$. Thus, one can also think of a functor as a generalized monotone map.

9. *An example from topology:* Let X be a topological space with collection of open sets $\mathcal{O}(X)$. Ordered by inclusion, $\mathcal{O}(X)$ is a poset category. Moreover, the points of X can be preordered by *specialization* by setting $x \leq y$ iff $x \in U$ implies $y \in U$ for every open set U, that is, y is contained in every open set that contains x. If X is sufficiently separated ("T_1"), then this ordering becomes trivial, but it can be quite interesting otherwise, as happens in the spaces of algebraic geometry and denotational semantics. It is an exercise to show that T_0 spaces are actually posets under the specialization ordering.

10. *An example from logic:* Given a deductive system of logic, there is an associated *category of proofs*, in which the objects are formulas:

$$\varphi, \psi, \ldots$$

An arrow from φ to ψ is a deduction of ψ from the (uncanceled) assumption φ.

$$\frac{\varphi}{\underset{\psi}{\vdots}}$$

Composition of arrows is given by putting together such deductions in the obvious way, which is clearly associative. (What should the identity arrows 1_φ be?) Observe that there can be many different arrows

$$p : \varphi \to \psi,$$

since there may be many different such proofs. This category turns out to have a very rich structure, which we consider later in connection with the λ-calculus.

11. *An example from computer science:* Given a functional programming language L, there is an associated category, where the objects are the data types of L, and the arrows are the computable functions of L ("processes," "procedures," "programs"). The composition of two such programs $X \xrightarrow{f} Y \xrightarrow{g} Z$ is given by applying g to the output of f, sometimes also

written as

$$g \circ f = f; g.$$

The identity is the "do nothing" program.

Categories such as this are basic to the idea of denotational semantics of programming languages. For example, if $\mathbf{C}(L)$ is the category just defined, then the denotational semantics of the language L in a category $\mathbf{D}$ of, say, Scott domains is simply a functor

$$S : \mathbf{C}(L) \to \mathbf{D}$$

since S assigns domains to the types of L and continuous functions to the programs. Both this example and the previous one are related to the notion of "cartesian closed category" that is considered later.

12. Let X be a set. We can regard X as a category $\mathbf{Dis}(X)$ by taking the objects to be the elements of X and taking the arrows to be just the required identity arrows, one for each $x \in X$. Such categories, in which the only arrows are identities, are called *discrete*. Note that discrete categories are just very special posets.

13. A *monoid* (sometimes called a *semigroup with unit*) is a set M equipped with a binary operation $\cdot : M \times M \to M$ and a distinguished "unit" element $u \in M$ such that for all $x, y, z \in M$,

$$x \cdot (y \cdot z) = (x \cdot y) \cdot z$$

and

$$u \cdot x = x = x \cdot u.$$

Equivalently, a monoid is a category with just one object. The arrows of the category are the elements of the monoid. In particular, the identity arrow is the unit element u. Composition of arrows is the binary operation $m \cdot n$ of the monoid.

Monoids are very common. There are the monoids of numbers like $\mathbb{N}$, $\mathbb{Q}$, or $\mathbb{R}$ with addition and 0, or multiplication and 1. But also for any set X, the set of functions from X to X, written as

$$\mathrm{Hom}_{\mathbf{Sets}}(X, X)$$

is a monoid under the operation of composition. More generally, for any object C in any category $\mathbf{C}$, the set of arrows from C to C, written as $\mathrm{Hom}_{\mathbf{C}}(C, C)$, is a monoid under the composition operation of $\mathbf{C}$.

Since monoids are structured sets, there is a category $\mathbf{Mon}$ whose objects are monoids and whose arrows are functions that preserve the monoid structure. In detail, a homomorphism from a monoid M to a monoid N is a function $h : M \to N$ such that for all $m, n \in M$,

$$h(m \cdot_M n) = h(m) \cdot_N h(n)$$

and

$$h(u_M) = u_N.$$

Observe that a monoid homomorphism from M to N is the same thing as a functor from M regarded as a category to N regarded as a category. In this sense, categories are also generalized monoids, and functors are generalized homomorphisms.

1.5 Isomorphisms

Definition 1.3. In any category **C**, an arrow $f : A \rightarrow B$ is called an *isomorphism*, if there is an arrow $g : B \rightarrow A$ in **C** such that

$$g \circ f = 1_A \quad \text{and} \quad f \circ g = 1_B.$$

Since inverses are unique (proof!), we write $g = f^{-1}$. We say that A is *isomorphic* to B, written $A \cong B$, if there exists an isomorphism between them.

The definition of isomorphism is our first example of an *abstract*, category theoretic definition of an important notion. It is abstract in the sense that it makes use only of the category theoretic notions, rather than some additional information about the objects and arrows. It has the advantage over other possible definitions that it applies in any category. For example, one sometimes defines an isomorphism of sets (monoids, etc.) as a *bijective* function (respectively, homomorphism), that is, one that is "1-1 and onto"—making use of the *elements* of the objects. This is *equivalent* to our definition in some cases, such as sets and monoids. But note that, for example in **Pos**, the category theoretic definition gives the right notion, while there are "bijective homomorphisms" between non-isomorphic posets. Moreover, in many cases *only* the abstract definition makes sense, as for example, in the case of a monoid regarded as a category.

Definition 1.4. A *group* G is a monoid with an inverse g^{-1} for every element g. Thus, G is a category with one object, in which every arrow is an isomorphism.

The natural numbers $\mathbb{N}$ do not form a group under either addition or multiplication, but the integers $\mathbb{Z}$ and the positive rationals $\mathbb{Q}^+$, respectively, do. For any set X, we have the group $\text{Aut}(X)$ of automorphisms (or "permutations") of X, that is, isomorphisms $f : X \rightarrow X$. (Why is this closed under "$\circ$"?) A *group of permutations* is a subgroup $G \subseteq \text{Aut}(X)$ for some set X, that is, a group of (some) automorphisms of X. Thus, the set G must satisfy the following:

1. The identity function 1_X on X is in G.
2. If $g, g' \in G$, then $g \circ g' \in G$.
3. If $g \in G$, then $g^{-1} \in G$.

A *homomorphism* of groups $h : G \to H$ is just a homomorphism of monoids, which then necessarily also preserves the inverses (proof!).

Now consider the following basic, classical result about abstract groups.

Theorem (Cayley). *Every group G is isomorphic to a group of permutations.*

Proof. (sketch)

1. First, define the Cayley representation $\bar{G}$ of G to be the following group of permutations of a set: the set is just G itself, and for each element $g \in G$, we have the permutation $\bar{g} : G \to G$, defined for all $h \in G$ by "acting on the left":

$$\bar{g}(h) = g \cdot h.$$

   This is indeed a permutation, since it has the action of g^{-1} as an inverse.
2. Next define homomorphisms $i : G \to \bar{G}$ by $i(g) = \bar{g}$, and $j : \bar{G} \to G$ by $j(\bar{g}) = \bar{g}(u)$.
3. Finally, show that $i \circ j = 1_{\bar{G}}$ and $j \circ i = 1_G$.

$\square$

Warning 1.5. Note the two different levels of isomorphisms that occur in the proof of Cayley's theorem. There are permutations of the set of elements of G, which are isomorphisms in **Sets**, and there is the isomorphism between G and $\bar{G}$, which is in the category **Groups** of groups and group homomorphisms.

Cayley's theorem says that any abstract group can be represented as a "concrete" one, that is, a group of permutations of a set. The theorem can in fact be generalized to show that any category that is not "too big" can be represented as one that is "concrete," that is, a category of sets and functions. (There is a technical sense of not being "too big" which is introduced in Section 1.8.)

Theorem 1.6. *Every category $\mathbf{C}$ with a set of arrows is isomorphic to one in which the objects are sets and the arrows are functions.*

Proof. (sketch) Define the Cayley representation $\bar{\mathbf{C}}$ of $\mathbf{C}$ to be the following concrete category:

- objects are sets of the form

$$\bar{C} = \{f \in \mathbf{C} \mid \operatorname{cod}(f) = C\}$$

  for all $C \in \mathbf{C}$,
- arrows are functions

$$\bar{g} : \bar{C} \to \bar{D}$$

for $g : C \to D$ in $\mathbf{C}$, defined for any $f : X \to C$ in $\bar{C}$ by $\bar{g}(f) = g \circ f$.

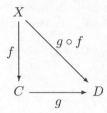

Remark 1.7. This shows us what is *wrong* with the naive notion of a "concrete" category of sets and functions: while not every category has special sets and functions as its objects and arrows, every category is isomorphic to such a one. Thus, the only special properties such categories can possess are ones that are categorically irrelevant, such as features of the objects that do not affect the arrows in any way (like the difference between the real numbers constructed as Dedekind cuts or as Cauchy sequences). A better attempt to capture what is intended by the rather vague idea of a "concrete" category is that arbitrary arrows $f : C \to D$ are completely determined by their composites with arrows $x : T \to C$ from some "test object" T, in the sense that $fx = gx$ for all such x implies $f = g$. As we shall see later, this amounts to considering a particular representation of the category, determined by T. A category is then said to be "concrete" when this condition holds for T a "terminal object," in the sense of Section 2.2; but there are also good reasons for considering other objects T, as we see Chapter 2.

Note that the condition that $\mathbf{C}$ has a *set* of arrows is needed to ensure that the collections $\{f \in \mathbf{C} \mid \mathrm{cod}(f) = C\}$ really are *sets*—we return to this point in Section 1.8.

1.6 Constructions on categories

Now that we have a stock of categories to work with, we can consider some constructions that produce new categories from old.

1. The *product* of two categories $\mathbf{C}$ and $\mathbf{D}$, written as

$$\mathbf{C} \times \mathbf{D}$$

has objects of the form (C, D), for $C \in \mathbf{C}$ and $D \in \mathbf{D}$, and arrows of the form

$$(f, g) : (C, D) \to (C', D')$$

for $f : C \to C' \in \mathbf{C}$ and $g : D \to D' \in \mathbf{D}$. Composition and units are defined componentwise, that is,

$$(f', g') \circ (f, g) = (f' \circ f, g' \circ g)$$
$$1_{(C,D)} = (1_C, 1_D).$$

There are two obvious *projection functors*

$$\mathbf{C} \xleftarrow{\;\;\pi_1\;\;} \mathbf{C} \times \mathbf{D} \xrightarrow{\;\;\pi_2\;\;} \mathbf{D}$$

defined by $\pi_1(C, D) = C$ and $\pi_1(f, g) = f$, and similarly for π_2.

The reader familiar with groups will recognize that for groups G and H, the product category $G \times H$ is the usual (direct) product of groups.

2. The *opposite* (or "dual") category $\mathbf{C}^{\mathrm{op}}$ of a category $\mathbf{C}$ has the same objects as $\mathbf{C}$, and an arrow $f : C \to D$ in $\mathbf{C}^{\mathrm{op}}$ is an arrow $f : D \to C$ in $\mathbf{C}$. That is, $\mathbf{C}^{\mathrm{op}}$ is just $\mathbf{C}$ with all of the arrows formally turned around.

It is convenient to have a notation to distinguish an object (resp. arrow) in $\mathbf{C}$ from the same one in $\mathbf{C}^{\mathrm{op}}$. Thus, let us write

$$f^* : D^* \to C^*$$

in $\mathbf{C}^{\mathrm{op}}$ for $f : C \to D$ in $\mathbf{C}$. With this notation, we can define composition and units in $\mathbf{C}^{\mathrm{op}}$ in terms of the corresponding operations in $\mathbf{C}$, namely,

$$1_{C^*} = (1_C)^*$$
$$f^* \circ g^* = (g \circ f)^*.$$

Thus, a diagram in $\mathbf{C}$

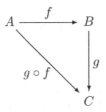

looks like this in $\mathbf{C}^{\mathrm{op}}$

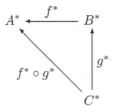

Many "duality" theorems of mathematics express the fact that one category is (a subcategory of) the opposite of another. An example of this sort which we prove later is that **Sets** is dual to the category of complete, atomic Boolean algebras.

3. The *arrow category* $\mathbf{C}^{\to}$ of a category $\mathbf{C}$ has the arrows of $\mathbf{C}$ as objects, and an arrow g from $f : A \to B$ to $f' : A' \to B'$ in $\mathbf{C}^{\to}$ is a "commutative"

square"

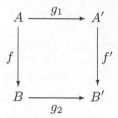

where g_1 and g_2 are arrows in **C**. That is, such an arrow is a pair of arrows $g = (g_1, g_2)$ in **C** such that

$$g_2 \circ f = f' \circ g_1.$$

The identity arrow 1_f on an object $f : A \to B$ is the pair $(1_A, 1_B)$. Composition of arrows is done componentwise:

$$(h_1, h_2) \circ (g_1, g_2) = (h_1 \circ g_1, h_2 \circ g_2)$$

The reader should verify that this works out by drawing the appropriate commutative diagram.

Observe that there are two functors:

$$\mathbf{C} \xleftarrow{\;\textbf{dom}\;} \mathbf{C}^{\to} \xrightarrow{\;\textbf{cod}\;} \mathbf{C}$$

4. The *slice category* **C**/C of a category **C** over an object $C \in \mathbf{C}$ has

- Objects: all arrows $f \in \mathbf{C}$ such that $\mathrm{cod}(f) = C$,
- Arrows: an arrow a from $f : X \to C$ to $f' : X' \to C$ is an arrow $a : X \to X'$ in **C** such that $f' \circ a = f$, as indicated in

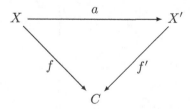

The identity arrows and composites are inherited from those of **C**, just as in the arrow category. Note that there is a functor $U : \mathbf{C}/C \to \mathbf{C}$ that "forgets about the base object C."

If $g : C \to D$ is any arrow, then there is a composition functor,

$$g_* : \mathbf{C}/C \to \mathbf{C}/D$$

defined by $g_*(f) = g \circ f$,

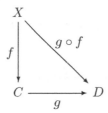

and similarly for arrows in $\mathbf{C}/C$. Indeed, the whole construction is a functor,

$$\mathbf{C}/(-) : \mathbf{C} \to \mathbf{Cat}$$

as the reader can easily verify. Compared to the Cayley representation, this functor gives a "representation" of $\mathbf{C}$ as a category of categories and functors—rather than sets and fuctions. Of course, the Cayley representation was just this one followed by the forgetful functor $U : \mathbf{Cat} \to \mathbf{Sets}$ that takes a category to its underlying set of objects.

If $\mathbf{C} = \mathbf{P}$ is a poset category and $p \in \mathbf{P}$, then

$$\mathbf{P}/p \cong \downarrow (p)$$

the slice category $\mathbf{P}/p$ is just the "principal ideal" $\downarrow (p)$ of elements $q \in \mathbf{P}$ with $q \leq p$. We will have more examples of slice categories soon.

The *coslice* category $C/\mathbf{C}$ of a category $\mathbf{C}$ under an object C of $\mathbf{C}$ has as objects all arrows f of $\mathbf{C}$ such that $\mathrm{dom}(f) = C$, and an arrow from $f : C \to X$ to $f' : C \to X'$ is an arrow $h : X \to X'$ such that $h \circ f = f'$. The reader should now carry out the rest of the definition of the coslice category by analogy with the definition of the slice category. How can the coslice category be defined in terms of the slice category and the opposite construction?

Example 1.8. The category $\mathbf{Sets}_*$ of *pointed sets* consists of sets A with a distinguished element $a \in A$, and arrows $f : (A, a) \to (B, b)$ are functions $f : A \to B$ that preserves the "points," $f(a) = b$. This is isomorphic to the coslice category,

$$\mathbf{Sets}_* \cong 1/\mathbf{Sets}$$

of Sets "under" any singleton $1 = \{*\}$. Indeed, functions $a : 1 \to A$ correspond uniquely to elements, $a(*) = a \in A$, and arrows $f : (A, a) \to (B, b)$ correspond exactly to commutative triangles:

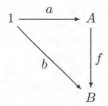

1.7 Free categories

Free monoid. Start with an "alphabet" A of "letters" $a, b, c, \ldots$, that is, a set,

$$A = \{a, b, c, \ldots\}.$$

A *word* over A is a finite sequence of letters:

$$thisword, \quad categoriesarefun, \quad asddjbnzzfj, \ldots$$

We write "-" for the empty word. The "Kleene closure" of A is defined to be the set

$$A^* = \{\text{words over } A\}.$$

Define a binary operation "$*$" on A^* by $w * w' = ww'$ for words $w, w' \in A^*$. Thus, "$*$" is just *concatenation*. The operation "$*$" is thus associative, and the empty word "-" is a unit. Thus, A^* is a monoid—called the *free monoid* on the set A. The elements $a \in A$ can be regarded as words of length one, so we have a function

$$i : A \to A^*$$

defined by $i(a) = a$, and called the "insertion of generators." The elements of A "generate" the free monoid, in the sense that every $w \in A^*$ is a $*$-product of a's, that is, $w = a_1 * a_2 * \cdots * a_n$ for some $a_1, a_2, ..., a_n$ in A.

Now what does "free" mean here? Any guesses? One sometimes sees definitions in "baby algebra" books along the following lines:

A monoid M is *freely generated* by a subset A of M, if the following conditions hold:

1. Every element $m \in M$ can be written as a product of elements of A:

$$m = a_1 \cdot_M \cdots \cdot_M a_n , \quad a_i \in A.$$

2. No "nontrivial" relations hold in M, that is, if $a_1 \ldots a_j = a'_1 \ldots a'_k$, then this is required by the axioms for monoids.

The first condition is sometimes called "no junk," while the second condition is sometimes called "no noise." Thus, the free monoid on A is a monoid containing A and having no junk and no noise. What do you think of this definition of a free monoid?

I would object to the reference in the second condition to "provability," or something. This must be made more precise for this to succeed as a definition. In category theory, we give a precise definition of "free"—capturing what is meant in the above—which avoids such vagueness.

First, every monoid N has an underlying set $|N|$, and every monoid homomorphism $f : N \to M$ has an underlying function $|f| : |N| \to |M|$. It is easy to see that this is a functor, called the "forgetful functor." The free monoid $M(A)$

on a set A is by definition "the" monoid with the following so called *universal mapping property* or UMP!

Universal Mapping Property of $M(A)$

There is a function $i : A \to |M(A)|$, and given any monoid N and any function $f : A \to |N|$, there is a *unique* monoid homomorphism $\bar{f} : M(A) \to N$ such that $|\bar{f}| \circ i = f$, all as indicated in the following diagram:

in **Mon**:

$$M(A) \quad \overset{\bar{f}}{\cdots\cdots\cdots\rightarrow} \quad N$$

in **Sets**:

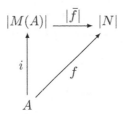

Proposition 1.9. *A^* has the UMP of the free monoid on A.*

Proof. Given $f : A \to |N|$, define $\bar{f} : A^* \to N$ by

$$\bar{f}(-) = u_N, \quad \text{the unit of } N$$
$$\bar{f}(a_1 \ldots a_i) = f(a_1) \cdot_N \ldots \cdot_N f(a_i).$$

Then, $\bar{f}$ is clearly a homomorphism with

$$\bar{f}(a) = f(a) \quad \text{for all } a \in A.$$

If $g : A^* \to N$ also satisfies $g(a) = f(a)$ for all $a \in A$, then for all $a_1 \ldots a_i \in A^*$:

$$\begin{aligned}
g(a_1 \ldots a_i) &= g(a_1 * \ldots * a_i) \\
&= g(a_1) \cdot_N \ldots \cdot_N g(a_i) \\
&= f(a_1) \cdot_N \ldots \cdot_N f(a_i) \\
&= \bar{f}(a_1) \cdot_N \ldots \cdot_N \bar{f}(a_i) \\
&= \bar{f}(a_1 * \ldots * a_i) \\
&= \bar{f}(a_1 \ldots a_i).
\end{aligned}$$

So, $g = \bar{f}$, as required.   □

Think about why the above UMP captures precisely what is meant by "no junk" and "no noise." Specifically, the existence part of the UMP captures the vague notion of "no noise" (because any equation that holds between algebraic combinations of the generators must also hold anywhere they can be mapped to, and thus everywhere), while the uniqueness part makes precise the "no junk" idea (because any extra elements not combined from the generators would be free to be mapped to *different* values).

Using the UMP, it is easy to show that the free monoid $M(A)$ is determined uniquely up to isomorphism, in the following sense.

Proposition 1.10. *Given monoids M and N with functions $i: A \to |M|$ and $j: A \to |N|$, each with the UMP of the free monoid on A, there is a (unique) monoid isomorphism $h: M \cong N$ such that $|h|i = j$ and $|h^{-1}|j = i$.*

Proof. From j and the UMP of M, we have $\bar{j}: M \to N$ with $|\bar{j}|i = j$ and from i and the UMP of N, we have $\bar{i}: N \to M$ with $|\bar{i}|j = i$. Composing gives a homomorphism $\bar{i} \circ \bar{j}: M \to M$ such that $|\bar{i} \circ \bar{j}|i = i$. Since $1_M: M \to M$ also has this property, by the uniqueness part of the UMP of M, we have $\bar{i} \circ \bar{j} = 1_M$. Exchanging the roles of M and N shows $\bar{j} \circ \bar{i} = 1_N$:

in **Mon***:*

$$M \xrightarrow{\quad \bar{j} \quad} N \xrightarrow{\quad \bar{i} \quad} M$$

in **Sets***:*

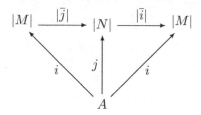

For example, the free monoid on any set with a single element is easily seen to be isomorphic to the monoid of natural numbers $\mathbb{N}$ under addition (the "generator" is the number 1). Thus, as a monoid, $\mathbb{N}$ is uniquely determined up to isomorphism by the UMP of free monoids.

Free category. Now, we want to do the same thing for categories in general (not just monoids). Instead of underlying sets, categories have underlying graphs, so let us review these first.

A *directed graph* consists of vertices and edges, each of which is directed, that is, each edge has a "source" and a "target" vertex.

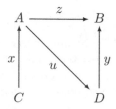

We draw graphs just like categories, but there is no composition of edges, and there are no identities.

A graph thus consists of two sets, E (edges) and V (vertices), and two functions, $s : E \to V$ (source) and $t : E \to V$ (target). Thus, in **Sets**, a graph is just a configuration of objects and arrows of the form

$$E \underset{t}{\overset{s}{\rightrightarrows}} V$$

Now, every graph G "generates" a category $\mathbf{C}(G)$, the *free category* on G. It is defined by taking the vertices of G as objects, and the *paths* in G as arrows, where a path is a finite sequence of edges $e_1, \ldots, e_n$ such that $t(e_i) = s(e_{i+1})$, for all $i = 1 \ldots n$. We write the arrows of $\mathbf{C}(G)$ in the form $e_n e_{n-1} \ldots e_1$.

$$v_0 \xrightarrow{e_1} v_1 \xrightarrow{e_2} v_2 \xrightarrow{e_3} \ldots \xrightarrow{e_n} v_n$$

Put

$$\mathrm{dom}(e_n \ldots e_1) = s(e_1)$$
$$\mathrm{cod}(e_n \ldots e_1) = t(e_n)$$

and define composition by concatenation:

$$e_n \ldots e_1 \circ e'_m \ldots e'_1 = e_n \ldots e_1 e'_m \ldots e'_1.$$

For each vertex v, we have an "empty path" denoted by 1_v, which is to be the identity arrow at v.

Note that if G has only one vertex, then $\mathbf{C}(G)$ is just the free monoid on the set of edges of G. Also note that if G has only vertices (no edges), then $\mathbf{C}(G)$ is the discrete category on the set of vertices of G.

Later on, we will have a general definition of "free." For now, let us see that $\mathbf{C}(G)$ also has a UMP. First, define a "forgetful functor"

$$U : \mathbf{Cat} \to \mathbf{Graphs}$$

in the obvious way: the underlying graph of a category $\mathbf{C}$ has as edges the arrows of $\mathbf{C}$, and as vertices the objects, with $s = \mathrm{dom}$ and $t = \mathrm{cod}$. The action of U on functors is equally clear, or at least it will be, once we have defined the arrows in **Graphs**.

A homomorphism of graphs is of course a "functor without the conditions on identities and composition," that is, a mapping of edges to edges and vertices to vertices that preserves sources and targets. We describe this from a slightly different point of view, which will be useful later on.

First, observe that we can describe a category $\mathbf{C}$ with a diagram like this:

$$C_2 \xrightarrow{\ \circ\ } C_1 \overset{\mathrm{cod}}{\underset{\mathrm{dom}}{\rightleftarrows}} \ \xleftarrow{\ i\ } C_0$$

where C_0 is the collection of objects of $\mathbf{C}$, C_1 the arrows, i is the identity arrow operation, and C_2 is the collection $\{(f, g) \in C_1 \times C_1 : \mathrm{cod}(f) = \mathrm{dom}(g)\}$.

Then a functor $F : \mathbf{C} \to \mathbf{D}$ from $\mathbf{C}$ to another category $\mathbf{D}$ is a pair of functions

$$F_0 : C_0 \to D_0$$
$$F_1 : C_1 \to D_1$$

such that each similarly labeled square in the following diagram commutes:

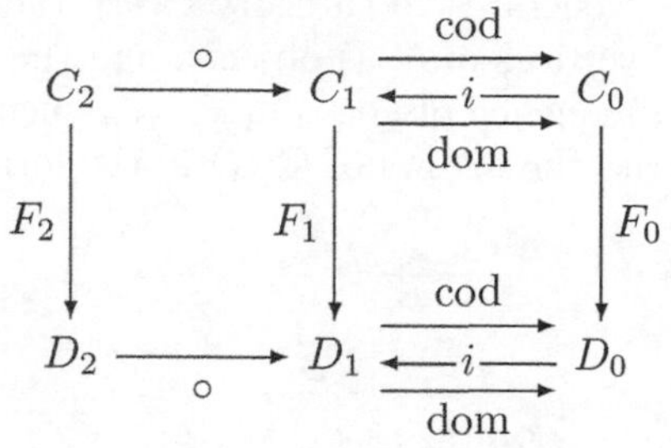

$$
\begin{array}{ccccc}
C_2 & \xrightarrow{\;\circ\;} & C_1 & \underset{\mathrm{dom}}{\overset{\mathrm{cod}}{\underset{\longleftarrow i}{\rightrightarrows}}} & C_0 \\
\Big\downarrow{F_2} & & \Big\downarrow{F_1} & & \Big\downarrow{F_0} \\
D_2 & \xrightarrow[\;\circ\;]{} & D_1 & \underset{\mathrm{dom}}{\overset{\mathrm{cod}}{\underset{\longleftarrow i}{\rightrightarrows}}} & D_0
\end{array}
$$

where $F_2(f, g) = (F_1(f), F_1(g))$.

Now let us describe a *homomorphism of graphs*,

$$h : G \to H.$$

We need a pair of functions $h_0 : G_0 \to H_0$, $h_1 : G_1 \to H_1$ making the two squares (once with t's, once with s's) in the following diagram commute:

$$
\begin{array}{ccc}
G_1 & \underset{s}{\overset{t}{\rightrightarrows}} & G_0 \\
\Big\downarrow{h_1} & & \Big\downarrow{h_0} \\
H_1 & \underset{s}{\overset{t}{\rightrightarrows}} & H_0
\end{array}
$$

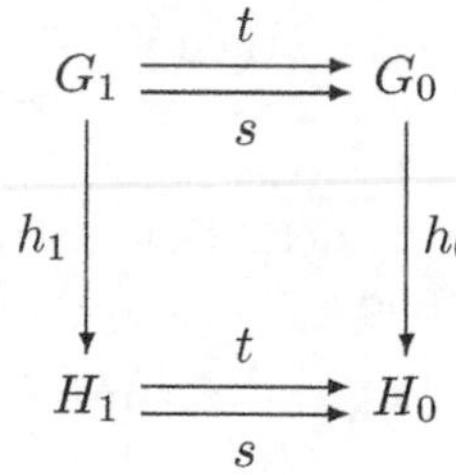

In these terms, we can easily describe the forgetful functor,

$$U : \mathbf{Cat} \to \mathbf{Graphs}$$

as sending the category

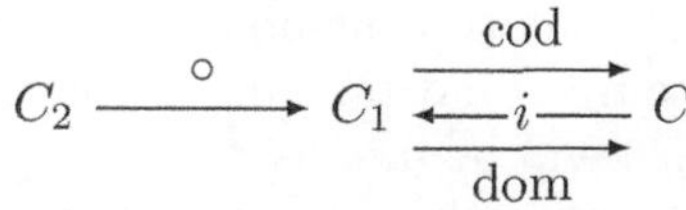

$$
C_2 \xrightarrow{\;\circ\;} C_1 \underset{\mathrm{dom}}{\overset{\mathrm{cod}}{\underset{\longleftarrow i}{\rightrightarrows}}} C_0
$$

to the underlying graph

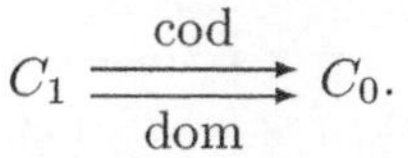

$$
C_1 \underset{\mathrm{dom}}{\overset{\mathrm{cod}}{\rightrightarrows}} C_0.
$$

And similarly for functors, the effect of U is described by simply erasing some parts of the diagrams (which is easier to demonstrate with chalk!). Let us again write $|\mathbf{C}| = U(\mathbf{C})$, etc., for the underlying graph of a category $\mathbf{C}$, in analogy to the case of monoids above.

The free category on a graph now has the following UMP.

Universal Mapping Property of $\mathbf{C}(G)$
There is a graph homomorphism $i : G \to |\mathbf{C}(G)|$, and given any category $\mathbf{D}$ and any graph homomorphism $h : G \to |\mathbf{D}|$, there is a *unique* functor $\bar{h} : \mathbf{C}(G) \to \mathbf{D}$ with $|\bar{h}| \circ i = h$.

in **Cat***:*

$$\mathbf{C}(G) \overset{\bar{h}}{\dashrightarrow} \mathbf{D}$$

in **Graph***:*

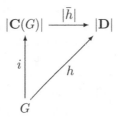

The free category on a graph with just one vertex is just a free monoid on the set of edges. The free category on a graph with two vertices and one edge between them is the finite category **2**. The free category on a graph of the form

$$A \underset{f}{\overset{e}{\rightleftarrows}} B$$

has (in addition to the identity arrows) the infinitely many arrows:

$$e, f, ef, fe, efe, fef, efef, \ldots$$

1.8 Foundations: large, small, and locally small

Let us begin by distinguishing between the following things:

(i) categorical foundations for mathematics,

(ii) mathematical foundations for category theory.

As for the first point, one sometimes hears it said that category theory can be used to provide "foundations for mathematics," as an alternative to set theory. That is in fact the case, but it is not what we are doing here. In set theory, one often begins with existential axioms such as "there is an infinite set" and derives further sets by axioms like "every set has a powerset," thus building up a universe of mathematical objects (namely sets), which in principle suffice for "all of mathematics." Our axiom that every arrow has a domain and a codomain is not to be understood in the same way as set theory's axiom that every set has a powerset! The difference is that in set theory—at least as usually conceived—the axioms are to be regarded as referring to (or determining) a single universe of sets. In category theory, by contrast, the axioms are a *definition* of something, namely of categories. This is just like in group theory or topology, where the axioms serve to define the objects under investigation. These, in turn, are assumed to exist in some "background" or "foundational" system, like set theory (or type theory). That theory of sets could itself, in turn, be determined using category theory, or in some other way.

This brings us to the second point: we assume that our categories are comprised of sets and functions, in one way or another, like most mathematical objects, and taking into account the remarks just made about the possibility of categorical (or other) foundations. But in category theory, we sometimes run into difficulties with set theory as usually practiced. Mostly these are questions of size; some categories are "too big" to be handled comfortably in conventional set theory. We already encountered this issue when we considered the Cayley representation in Section 1.5. There we had to require that the category under consideration had (no more than) a set of arrows. We would certainly not want to impose this restriction in general, however (as one usually does for, say, groups); for then even the "category" **Sets** would fail to be a proper category, as would many other categories that we definitely want to study.

There are various formal devices for addressing these issues, and they are discussed in the book by Mac Lane. For our immediate purposes, the following distinction will be useful.

Definition 1.11. A category $\mathbf{C}$ is called *small* if both the collection $\mathbf{C}_0$ of objects of $\mathbf{C}$ and the collection $\mathbf{C}_1$ of arrows of $\mathbf{C}$ are sets. Otherwise, $\mathbf{C}$ is called *large*.

For example, all finite categories are clearly small, as is the category **Sets**$_{\text{fin}}$ of finite sets and functions. (Actually, one should stipulate that the sets are only built from other finite sets, all the way down, i.e., that they are "hereditarily finite".) On the other hand, the category **Pos** of posets, the category **Groups** of groups, and the category **Sets** of sets are all large. We let **Cat** be the category of all *small categories*, which itself is a large category. In particular, then, **Cat** is not an object of itself, which may come as a relief to some readers.

This does not really solve all of our difficulties. Even for large categories like **Groups** and **Sets** we will want to also consider constructions like the category of all functors from one to the other (we define this "functor category" later). But if these are not small, conventional set theory does not provide the means to do this directly (these categories would be "too large"). Therefore, one needs a more elaborate theory of "classes" to handle such constructions. We will not worry about this when it is just a matter of technical foundations (Mac Lane I.6 addresses this issue). However, one very useful notion in this connection is the following.

Definition 1.12. A category $\mathbf{C}$ is called *locally small* if for all objects X, Y in $\mathbf{C}$, the collection $\mathrm{Hom}_{\mathbf{C}}(X, Y) = \{ f \in \mathbf{C}_1 \mid f : X \to Y \}$ is a *set* (called a *hom-set*).

Many of the large categories we want to consider are, in fact, locally small. **Sets** is locally small since $\mathrm{Hom}_{\mathbf{Sets}}(X, Y) = Y^X$, the *set* of all functions from X to Y. Similarly, **Pos**, **Top**, and **Group** are all locally small (is **Cat**?), and, of course, any small category is locally small.

Warning 1.13. Do not confuse the notions *concrete* and *small*. To say that a category is concrete is to say that the *objects* of the category are (structured) sets, and the arrows of the category are (certain) functions. To say that a category is small is to say that the *collection of all objects* of the category is a set, as is the collection of all arrows. The real numbers $\mathbb{R}$, regarded as a poset category, is small but not concrete. The category **Pos** of all posets is concrete but not small.

1.9   Exercises

1. The objects of **Rel** are sets, and an arrow $A \to B$ is a relation from A to B, that is, a subset $R \subseteq A \times B$. The equality relation $\{ \langle a, a \rangle \in A \times A \mid a \in A \}$ is the identity arrow on a set A. Composition in **Rel** is to be given by

$$S \circ R = \{ \langle a, c \rangle \in A \times C \mid \exists b \, (\langle a, b \rangle \in R \ \& \ \langle b, c \rangle \in S) \}$$

   for $R \subseteq A \times B$ and $S \subseteq B \times C$.

   (a) Show that **Rel** is a category.

   (b) Show also that there is a functor $G : \mathbf{Sets} \to \mathbf{Rel}$ taking objects to themselves and each function $f : A \to B$ to its graph,

$$G(f) = \{ \langle a, f(a) \rangle \in A \times B \mid a \in A \}.$$

   (c) Finally, show that there is a functor $C : \mathbf{Rel}^{\mathrm{op}} \to \mathbf{Rel}$ taking each relation $R \subseteq A \times B$ to its converse $R^c \subseteq B \times A$, where,

$$\langle a, b \rangle \in R^c \Leftrightarrow \langle b, a \rangle \in R.$$

2. Consider the following isomorphisms of categories and determine which hold.

   (a) $\mathbf{Rel} \cong \mathbf{Rel}^{\mathrm{op}}$

   (b) $\mathbf{Sets} \cong \mathbf{Sets}^{\mathrm{op}}$

(c) For a fixed set X with powerset $P(X)$, as poset categories $P(X) \cong P(X)^{\text{op}}$ (the arrows in $P(X)$ are subset inclusions $A \subseteq B$ for all $A, B \subseteq X$).

3. (a) Show that in **Sets**, the isomorphisms are exactly the bijections.

   (b) Show that in **Monoids**, the isomorphisms are exactly the bijective homomorphisms.

   (c) Show that in **Posets**, the isomorphisms are *not* the same as the bijective homomorphisms.

4. Let X be a topological space and preorder the points by *specialization*: $x \leq y$ iff y is contained in every open set that contains x. Show that this is a preorder, and that it is a poset if X is T_0 (for any two distinct points, there is some open set containing one but not the other). Show that the ordering is trivial if X is T_1 (for any two distinct points, each is contained in an open set not containing the other).

5. For any category $\mathbf{C}$, define a functor $U : \mathbf{C}/C \to \mathbf{C}$ from the slice category over an object C that "forgets about C." Find a functor $F : \mathbf{C}/C \to \mathbf{C}^{\to}$ to the arrow category such that $\mathbf{dom} \circ F = U$.

6. Construct the "coslice category" $C/\mathbf{C}$ of a category $\mathbf{C}$ under an object C from the slice category $\mathbf{C}/C$ and the "dual category" operation $-^{\text{op}}$.

7. Let $2 = \{a, b\}$ be any set with exactly 2 elements a and b. Define a functor $F : \mathbf{Sets}/2 \to \mathbf{Sets} \times \mathbf{Sets}$ with $F(f : X \to 2) = (f^{-1}(a), f^{-1}(b))$. Is this an isomorphism of categories? What about the analogous situation with a one-element set $1 = \{a\}$ instead of 2?

8. Any category $\mathbf{C}$ determines a preorder $P(\mathbf{C})$ by defining a binary relation $\leq$ on the objects by

$$A \leq B \text{ if and only if there is an arrow } A \to B$$

Show that P determines a functor from categories to preorders, by defining its effect on functors between categories and checking the required conditions. Show that P is a (one-sided) inverse to the evident inclusion functor of preorders into categories.

9. Describe the free categories on the following graphs by determining their objects, arrows, and composition operations.

(a)

$$a \xrightarrow{\;\;e\;\;} b$$

(b)

$$a \underset{f}{\overset{e}{\rightleftarrows}} b$$

(c)

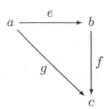

(d)

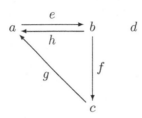

10. How many free categories on graphs are there which have exactly six arrows? Draw the graphs that generate these categories.

11. Show that the free monoid functor

$$M : \mathbf{Sets} \to \mathbf{Mon}$$

exists, in two different ways:

(a) Assume the particular choice $M(X) = X^*$ and define its effect

$$M(f) : M(A) \to M(B)$$

on a function $f : A \to B$ to be

$$M(f)(a_1 \dots a_k) = f(a_1) \dots f(a_k), \quad a_1, \dots a_k \in A.$$

(b) Assume only the UMP of the free monoid and use it to determine M on functions, showing the result to be a functor.

Reflect on how these two approaches are related.

12. Verify the UMP for free categories on graphs, defined as above with arrows being sequences of edges. Specifically, let $\mathbf{C}(G)$ be the free category on the graph G, so defined, and $i : G \to U(\mathbf{C}(G))$ the graph homomorphism taking vertices and edges to themselves, regarded as objects and arrows in $\mathbf{C}(G)$. Show that for any category $\mathbf{D}$ and graph homomorphism $f : G \to U(\mathbf{D})$, there is a unique functor

$$\bar{h} : \mathbf{C}(G) \to \mathbf{D}$$

with

$$U(\bar{h}) \circ i = h,$$

where $U : \mathbf{Cat} \to \mathbf{Graph}$ is the underlying graph functor.

13. Use the Cayley representation to show that every small category is iso-morphic to a "concrete" one, that is, one in which the objects are sets and the arrows are functions between them.

14. The notion of a category can also be defined with just one sort (arrows) rather than two (arrows and objects); the domains and codomains are taken to be certain *arrows* that act as units under composition, which is partially defined. Read about this definition in section I.1 of Mac Lane's *Categories for the Working Mathematician*, and do the exercise mentioned there, showing that it is equivalent to the usual definition.

2

ABSTRACT STRUCTURES

We begin with some remarks about category-theoretical definitions. These are characterizations of properties of objects and arrows in a category solely in terms of other objects and arrows, that is, just in the language of category theory. Such definitions may be said to be abstract, structural, operational, relational, or perhaps external (as opposed to internal). The idea is that objects and arrows are determined by the role they play in the category via their relations to other objects and arrows, that is, by their position in a structure and not by what they "are" or "are made of" in some absolute sense. The free monoid or category construction of the foregoing chapter was an example of one such definition, and we see many more examples of this kind later; for now, we start with some very simple ones. Let us call them *abstract characterizations*. We see that one of the basic ways of giving such an abstract characterization is via a Universal Mapping Property (UMP).

2.1 Epis and monos

Recall that in **Sets**, a function $f : A \to B$ is called

injective if $f(a) = f(a')$ implies $a = a'$ for all $a, a' \in A$,
surjective if for all $b \in B$ there is some $a \in A$ with $f(a) = b$.

We have the following abstract characterizations of these properties.

Definition 2.1. In any category **C**, an arrow

$$f : A \to B$$

is called a

monomorphism, if given any $g, h : C \to A$, $fg = fh$ implies $g = h$,

$$C \underset{h}{\overset{g}{\rightrightarrows}} A \xrightarrow{\;f\;} B$$

epimorphism, if given any $i, j : B \to D$, $if = jf$ implies $i = j$,

$$A \xrightarrow{\;f\;} B \underset{j}{\overset{i}{\rightrightarrows}} D.$$

We often write $f : A \rightarrowtail B$ if f is a monomorphism and $f : A \twoheadrightarrow B$ if f is an epimorphism.

Proposition 2.2. *A function $f : A \to B$ between sets is monic just in case it is injective.*

Proof. Suppose $f : A \rightarrowtail B$. Let $a, a' \in A$ such that $a \neq a'$, and let $\{x\}$ be any given one-element set. Consider the functions

$$\bar{a}, \bar{a'} : \{x\} \to A$$

where

$$\bar{a}(x) = a, \qquad \bar{a'}(x) = a'.$$

Since $\bar{a} \neq \bar{a'}$, it follows, since f is a monomorphism, that $f\bar{a} \neq f\bar{a'}$. Thus, $f(a) = (f\bar{a})(x) \neq (f\bar{a'})(x) = f(a')$. Whence f is injective.

Conversely, if f is injective and $g, h : C \to A$ are functions such that $g \neq h$, then for some $c \in C$, $g(c) \neq h(c)$. Since f is injective, it follows that $f(g(c)) \neq f(h(c))$, whence $fg \neq fh$. □

Example 2.3. In many categories of "structured sets" like monoids, the monos are exactly the "injective homomorphisms." More precisely, a homomorphism $h : M \to N$ of monoids is monic just if the underlying function $|h| : |M| \to |N|$ is monic, that is, injective by the foregoing. To prove this, suppose h is monic and take two different "elements" $x, y : 1 \to |M|$, where $1 = \{*\}$ is any one-element set. By the UMP of the free monoid $M(1)$ there are distinct corresponding homomorphisms $\bar{x}, \bar{y} : M(1) \to M$, with distinct composites $h \circ \bar{x}, h \circ \bar{y} : M(1) \to M \to N$, since h is monic. Thus, the corresponding "elements" $hx, hy : 1 \to N$ of N are also distinct, again by the UMP of $M(1)$.

$$M(1) \underset{\bar{y}}{\overset{\bar{x}}{\rightrightarrows}} M \xrightarrow{\;h\;} N$$

$$1 \underset{y}{\overset{x}{\rightrightarrows}} |M| \xrightarrow{\;|h|\;} |N|$$

Conversely, if $|h| : |M| \to |N|$ is monic and $f, g : X \to M$ are any distinct homomorphisms, then $|f|, |g| : |X| \to |M|$ are distinct functions, and so $|h| \circ |f|, |h| \circ |g| : |X| \to |M| \to |N|$ are distinct, since $|h|$ is monic. Since therefore $|h \circ f| = |h| \circ |f| \neq |h| \circ |g| = |h \circ g|$, we also must have $h \circ f \neq h \circ g$.

A completely analogous situation holds, for example, for groups, rings, vector spaces, and posets. We shall see that this fact follows from the presence, in each of these categories, of certain objects like the free monoid $M(1)$.

Example 2.4. In a poset $\mathbf{P}$, every arrow $p \leq q$ is both monic and epic. Why?

Now, dually to the foregoing, the epis in **Sets** are exactly the surjective functions (exercise!); by contrast, however, in many other familiar categories they are not just the surjective homomorphisms, as the following example shows.

Example 2.5. In the category **Mon** of monoids and monoid homomorphisms, there is a monic homomorphism

$$\mathbb{N} \rightarrowtail \mathbb{Z}$$

where $\mathbb{N}$ is the additive monoid $(N, +, 0)$ of natural numbers and $\mathbb{Z}$ is the additive monoid $(Z, +, 0)$ of integers. We show that this map, given by the inclusion $N \subset Z$ of sets, is also epic in **Mon** by showing that the following holds:

Given any monoid homomorphisms $f, g : (\mathbb{Z}, +, 0) \rightarrow (M, *, u)$, if the restrictions to N are equal, $f \mid_N = g \mid_N$, then $f = g$.

Note first that

$$f(-n) = f((-1)_1 + (-1)_2 + \cdots + (-1)_n)$$
$$= f(-1)_1 * f(-1)_2 * \cdots * f(-1)_n$$

and similarly for g. It, therefore, suffices to show that $f(-1) = g(-1)$. But

$$\begin{aligned}
f(-1) &= f(-1) * u \\
&= f(-1) * g(0) \\
&= f(-1) * g(1-1) \\
&= f(-1) * g(1) * g(-1) \\
&= f(-1) * f(1) * g(-1) \\
&= f(-1+1) * g(-1) \\
&= f(0) * g(-1) \\
&= u * g(-1) \\
&= g(-1).
\end{aligned}$$

Note that, from an algebraic point of view, a morphism e is epic if and only if e cancels on the right: $xe = ye$ implies $x = y$. Dually, m is monic if and only if m cancels on the left: $mx = my$ implies $x = y$.

Proposition 2.6. *Every iso is both monic and epic.*

Proof. Consider the following diagram:

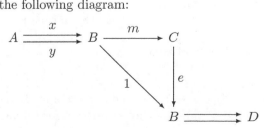

If m is an isomorphism with inverse e, then $mx = my$ implies $x = emx = emy = y$. Thus, m is monic. Similarly, e cancels on the right and thus is epic. $\square$

In **Sets**, the converse of the foregoing also holds: every mono-epi is iso. But this is not in general true, as shown by the example in monoids above.

2.1.1 Sections and retractions

We have just noted that any iso is both monic and epic. More generally, if an arrow

$$f : A \to B$$

has a left inverse

$$g : B \to A, \quad gf = 1_A$$

then f must be monic and g epic, by the same argument.

Definition 2.7. A *split* mono (epi) is an arrow with a left (right) inverse. Given arrows $e : X \to A$ and $s : A \to X$ such that $es = 1_A$, the arrow s is called a *section* or *splitting* of e, and the arrow e is called a *retraction* of s. The object A is called a *retract* of X.

Since functors preserve identities, they also preserve *split* epis and *split* monos. Compare example 2.5 above in **Mon** where the forgetful functor

$$\textbf{Mon} \to \textbf{Set}$$

did not preserve the epi $\mathbb{N} \to \mathbb{Z}$.

Example 2.8. In **Sets**, every mono splits except those of the form

$$\emptyset \rightarrowtail A.$$

The condition that *every epi splits* is the categorical version of the axiom of choice. Indeed, consider an epi

$$e : E \twoheadrightarrow X.$$

We have the family of nonempty sets:

$$E_x = e^{-1}\{x\}, \quad x \in X.$$

A choice function for this family $(E_x)_{x \in X}$ is exactly a splitting of e, that is, a function $s : X \to E$ such that $es = 1_X$, since this means that $s(x) \in E_x$ for all $x \in X$.

Conversely, given a family of nonempty sets,

$$(E_x)_{x \in X}$$

take $E = \{(x,y) \mid x \in X,\ y \in E_x\}$ and define the epi $e : E \twoheadrightarrow X$ by $(x,y) \mapsto x$. A splitting s of e then determines a choice function for the family.

The idea that a "family of objects" $(E_x)_{x \in X}$ can be represented by a single arrow $e : E \to X$ by using the "fibers" $e^{-1}(x)$ has much wider application than this, and is considered further in Section 7.10.

A notion related to the existence of "choice functions" is that of being "projective": an object P is said to be *projective* if for any epi $c : E \twoheadrightarrow X$ and arrow $f : P \to X$ there is some (not necessarily unique) arrow $\bar{f} : P \to E$ such that $e \circ \bar{f} = f$, as indicated in the following diagram:

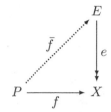

One says that f *lifts across* e. Any epi into a projective object clearly splits. Projective objects may be thought of as having a more "free" structure, thus permitting "more arrows."

The axiom of choice implies that all sets are projective, and it follows that free objects in many (but not all!) categories of algebras then are also projective. The reader should show that, in any category, any retract of a projective object is also projective.

2.2 Initial and terminal objects

We now consider abstract characterizations of the empty set and the one-element sets in the category **Sets** and structurally similar objects in general categories.

Definition 2.9. In any category **C**, an object

0 is *initial* if for any object C there is a unique morphism

$$0 \to C,$$

1 is *terminal* if for any object C there is a unique morphism

$$C \to 1.$$

As in the case of monos and epis, note that there is a kind of "duality" in these definitions. Precisely, a terminal object in **C** is exactly an initial object in **C**^{op}. We consider this duality systematically in Chapter 3.

First, observe that since the notions of initial and terminal object are simple UMPs, such objects are uniquely determined up to isomorphism, just like the free monoids were.

Proposition 2.10. *Initial (terminal) objects are unique up to isomorphism.*

Proof. In fact, if C and C' are both initial (terminal) in the same category, then there is a *unique* isomorphism $C \to C'$. Indeed, suppose that 0 and $0'$ are both initial objects in some category $\mathbf{C}$; the following diagram then makes it clear that 0 and $0'$ are uniquely isomorphic:

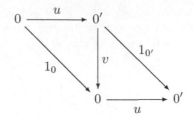

For terminal objects, apply the foregoing to $\mathbf{C}^{\mathrm{op}}$.   □

Example 2.11.

1. In **Sets**, the empty set is initial and any singleton set $\{x\}$ is terminal. Observe that **Sets** has just one initial object but many terminal objects (answering the question of whether **Sets** $\cong$ **Sets**^{op}).

2. In **Cat**, the category **0** (no objects and no arrows) is initial and the category **1** (one object and its identity arrow) is terminal.

3. In **Groups**, the one-element group is *both* initial and terminal (similarly for the category of vector spaces and linear transformations, as well as the category of monoids and monoid homomorphisms). But in **Rings** (commutative with unit), the ring $\mathbb{Z}$ of integers is initial (the one-element ring with $0 = 1$ is terminal).

4. A *Boolean algebra* is a poset B equipped with distinguished elements $0, 1$, binary operations $a \vee b$ of "join" and $a \wedge b$ of "meet," and a unary operation $\neg b$ of "complementation." These are required to satisfy the conditions

$$0 \leq a$$
$$a \leq 1$$
$$a \leq c \quad \text{and} \quad b \leq c \quad \text{iff} \quad a \vee b \leq c$$
$$c \leq a \quad \text{and} \quad c \leq b \quad \text{iff} \quad c \leq a \wedge b$$
$$a \leq \neg b \quad \text{iff} \quad a \wedge b = 0$$
$$\neg\neg a = a.$$

There is also an equivalent, fully equational characterization not involving the ordering. A typical example of a Boolean algebra is the powerset $\mathcal{P}(X)$ of all subsets $A \subseteq X$ of a set X, ordered by inclusion $A \subseteq B$, and with the Boolean operations being the empty set $0 = \emptyset$, the whole set $1 = X$, union and intersection of subsets as join and meet, and the relative complement $X - A$ as $\neg A$. A familiar special case is the two-element

Boolean algebra $\mathbf{2} = \{0, 1\}$ (which may be taken to be the powerset $\mathcal{P}(1)$), sometimes also regarded as "truth values" with the logical operations of disjunction, conjunction, and negation as the Boolean operations. It is an initial object in the category $\mathbf{BA}$ of Boolean algebras. $\mathbf{BA}$ has as arrows the Boolean homomorphisms, that is, functors $h : B \rightarrow B'$ that preserve the additional structure, in the sense that $h(0) = 0$, $h(a \vee b) = h(a) \vee h(b)$, etc. The one-element Boolean algebra (i.e., $\mathcal{P}(0)$) is terminal.

5. In a poset, an object is plainly initial iff it is the least element, and terminal iff it is the greatest element. Thus, for instance, any Boolean algebra has both. Obviously, a category *need not* have either an initial object or a terminal object; for example, the poset $(\mathbb{Z}, \leq)$ has neither.

6. For any category $\mathbf{C}$ and any object $X \in \mathbf{C}$, the identity arrow $1_X : X \rightarrow X$ is a terminal object in the slice category $\mathbf{C}/X$ and an initial object in the coslice category $X/\mathbf{C}$.

2.3 Generalized elements

Let us consider arrows into and out of initial and terminal objects. Clearly only certain of these will be of interest, but those are often especially significant.

A set A has an arrow into the initial object $A \rightarrow 0$ just if it is itself initial, and the same is true for posets. In monoids and groups, by contrast, every object has a unique arrow to the initial object, since it is also terminal.

In the category $\mathbf{BA}$ of Boolean algebras, however, the situation is quite different. The maps $p : B \rightarrow \mathbf{2}$ into the initial Boolean algebra $\mathbf{2}$ correspond uniquely to the so-called *ultrafilters* U in B. A *filter* in a Boolean algebra B is a nonempty subset $F \subseteq B$ that is closed upward and under meets:

$$a \in F \text{ and } a \leq b \quad \text{implies} \quad b \in F$$
$$a \in F \text{ and } b \in F \quad \text{implies} \quad a \wedge b \in F$$

A filter F is *maximal* if the only strictly larger filter $F \subset F'$ is the "improper" filter, namely all of B. An *ultrafilter* is a maximal filter. It is not hard to see that a filter F is an ultrafilter just if for every element $b \in B$, either $b \in F$ or $\neg b \in F$, and not both (exercise!). Now if $p : B \rightarrow \mathbf{2}$, let $U_p = p^{-1}(1)$ to get an ultrafilter $U_p \subset B$. And given an ultrafilter $U \subset B$, define $p_U(b) = 1$ iff $b \in U$ to get a Boolean homomorphism $p_U : B \rightarrow \mathbf{2}$. This is easy to check, as is the fact that these operations are mutually inverse. Boolean homomorphisms $B \rightarrow \mathbf{2}$ are also used in forming the "truth tables" one meets in logic. Indeed, a row of a truth table corresponds to such a homomorphism on a Boolean algebra of formulas.

Ring homomorphisms $A \rightarrow \mathbb{Z}$ into the initial ring $\mathbb{Z}$ play an analogous and equally important role in algebraic geometry. They correspond to so-called *prime ideals*, which are the ring-theoretic generalizations of ultrafilters.

Now let us consider some arrows from terminal objects. For any set X, for instance, we have an isomorphism

$$X \cong \mathrm{Hom}_{\mathbf{Sets}}(1, X)$$

between elements $x \in X$ and arrows $\bar{x} : 1 \to X$, determined by $\bar{x}(*) = x$, from a terminal object $1 = \{*\}$. We have already used this correspondence several times. A similar situation holds in posets (and in topological spaces), where the arrows $1 \to P$ correspond to elements of the underlying set of a poset (or space) P. In any category with a terminal object 1, such arrows $1 \to A$ are often called *global elements*, or *points*, or *constants* of A. In sets, posets, and spaces, the general arrows $A \to B$ are determined by what they do to the points of A, in the sense that two arrows $f, g : A \to B$ are equal if for every point $a : 1 \to A$ the composites are equal, $fa = ga$.

But be careful; this is not always the case! How many points are there of an object M in the category of monoids? That is, how many arrows of the form $1 \to M$ for a given monoid M? Just one! And how many points does a Boolean algebra have?

Because, in general, an object is not determined by its points, it is convenient to introduce the device of *generalized elements*. These are arbitrary arrows,

$$x : X \to A$$

(with arbitrary domain X), which can be regarded as *generalized* or *variable elements* of A. Computer scientists and logicians sometimes think of arrows $1 \to A$ as constants or closed terms and general arrows $X \to A$ as arbitrary terms. Summarizing:

Example 2.12.

1. Consider arrows $f, g : P \to Q$ in **Pos**. Then $f = g$ iff for all $x : 1 \to P$, we have $fx = gx$. In this sense, posets "have enough points" to separate the arrows.

2. By contrast, in **Mon**, for homomorphisms $h, j : M \to N$, we always have $hx = jx$, for all $x : 1 \to M$, since there is just one such point x. Thus, monoids do not "have enough points."

3. But in any category **C**, and for any arrows $f, g : C \to D$, we always have $f = g$ iff for all $x : X \to C$, it holds that $fx = gx$ (why?). Thus, all objects have enough generalized elements.

4. In fact, it often happens that it is enough to consider generalized elements of just a certain form $T \to A$, that is, for certain "test" objects T. We shall consider this presently.

Generalized elements are also good for "testing" for various conditions. Consider, for instance, diagrams of the following shape:

$$X \overset{x}{\underset{x'}{\rightrightarrows}} A \overset{f}{\longrightarrow} B$$

The arrow f is monic iff $x \neq x'$ implies $fx \neq fx'$ for all x, x', that is, just if f is "injective on generalized elements."

Similarly, in any category $\mathbf{C}$, to test whether a square commutes,

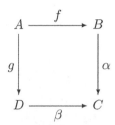

we shall have $\alpha f = \beta g$ just if $\alpha f x = \beta g x$ for all generalized elements $x : X \to A$ (just take $x = 1_A : A \to A$).

Example 2.13. Generalized elements can be used to "reveal more structure" than do the constant elements. For example, consider the following posets X and A:

$$X = \{x \leq y, x \leq z\}$$
$$A = \{a \leq b \leq c\}$$

There is an order-preserving bijection $f : X \to A$ defined by

$$f(x) = a, \qquad f(y) = b, \qquad f(z) = c.$$

It is easy to see that f is both monic and epic in the category **Pos**; however, it is clearly not an iso. One would like to say that X and A are "different structures," and indeed, their being nonisomorphic says just this. But now, how to *prove* that they are *not* isomorphic (say, via some other $X \to A$)? In general, this sort of thing can be quite difficult.

One way to prove that two objects are not isomorphic is to use "invariants": attributes that are preserved by isomorphisms. If two objects differ by an invariant they cannot be isomorphic. Generalized elements provide an easy way to define invariants. For instance, the number of global elements of X and A is the same, namely the three elements of the sets. But consider instead the "2-elements" $\mathbf{2} \to X$, from the poset $\mathbf{2} = \{0 \leq 1\}$ as a "test-object." Then X has 5 such elements, and A has 6. Since these numbers are invariants, the posets cannot be isomorphic. In more detail, we can define for any poset P the numerical invariant

$$|\mathrm{Hom}(\mathbf{2}, P)| \quad = \quad \text{the number of elements of } \mathrm{Hom}(\mathbf{2}, P).$$

Then if $P \cong Q$, it is easy to see that $|\operatorname{Hom}(\mathbf{2}, P)| = |\operatorname{Hom}(\mathbf{2}, Q)|$, since any isomorphism

$$P \underset{j}{\overset{i}{\rightleftarrows}} Q$$

also gives an iso

$$\operatorname{Hom}(\mathbf{2}, P) \underset{j_*}{\overset{i_*}{\rightleftarrows}} \operatorname{Hom}(\mathbf{2}, Q)$$

defined by composition:

$$i_*(f) = if$$
$$j_*(g) = jg$$

for all $f : \mathbf{2} \to P$ and $g : \mathbf{2} \to Q$. Indeed, this is a special case of the very general fact that $\operatorname{Hom}(X, -)$ is always a functor, and functors always preserve isos.

Example 2.14. As in the foregoing example, it is often the case that generalized elements $t : T \to A$ "based at" certain objects T are especially "revealing." We can think of such elements geometrically as "figures of shape T in A," just as an arrow $\mathbf{2} \to P$ in posets is a figure of shape $p \leq p'$ in P. For instance, as we have already seen, in the category of monoids, the arrows from the terminal monoid are entirely uninformative, but those from the free monoid on one generator $M(1)$ suffice to distinguish homomorphisms, in the sense that two homomorphisms $f, g : M \to M'$ are equal if their composites with all such arrows are equal. Since we know that $M(1) = \mathbb{N}$, the monoid of natural numbers, we can think of generalized elements $M(1) \to M$ based at $M(1)$ as "figures of shape $\mathbb{N}$" in M. In fact, by the UMP of $M(1)$, the underlying set $|M|$ is therefore (isomorphic to) the collection $\operatorname{Hom}_{\mathbf{Mon}}(\mathbb{N}, M)$ of all such figures, since

$$|M| \cong \operatorname{Hom}_{\mathbf{Sets}}(1, |M|) \cong \operatorname{Hom}_{\mathbf{Mon}}(\mathbb{N}, M).$$

In this sense, a map from a monoid is determined by its effect on all of the figures of shape $\mathbb{N}$ in the monoid.

2.4   Products

Next, we are going to see the categorical definition of a product of two objects in a category. This was first given by Mac Lane in 1950, and it is probably the earliest example of category theory being used to define a fundamental mathematical notion.

By "define" here I mean an abstract characterization, in the sense already used, in terms of objects and arrows in a category. And as before, we do this by giving a UMP, which determines the structure at issue up to isomorphism, as usual in category theory. Later in this chapter, we have several other examples of such characterizations.

Let us begin by considering products of sets. Given sets A and B, the *cartesian product* of A and B is the set of ordered pairs

$$A \times B = \{(a,b) \mid a \in A, \ b \in B\}.$$

Observe that there are two "coordinate projections"

$$A \xleftarrow{\ \pi_1\ } A \times B \xrightarrow{\ \pi_2\ } B$$

with

$$\pi_1(a,b) = a, \qquad \pi_2(a,b) = b.$$

And indeed, given any element $c \in A \times B$, we have

$$c = (\pi_1 c, \pi_2 c).$$

The situation is captured concisely in the following diagram:

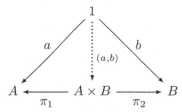

Replacing elements by generalized elements, we get the following definition.

Definition 2.15. In any category $\mathbf{C}$, a *product diagram* for the objects A and B consists of an object P and arrows

$$A \xleftarrow{\ p_1\ } P \xrightarrow{\ p_2\ } B$$

satisfying the following UMP:

Given any diagram of the form

$$A \xleftarrow{\ x_1\ } X \xrightarrow{\ x_2\ } B$$

there exists a unique $u : X \to P$, making the diagram

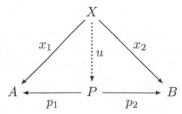

commute, that is, such that $x_1 = p_1 u$ and $x_2 = p_2 u$.

Remark 2.16. As in other UMPs, there are two parts:

Existence: There is some $u : X \to U$ such that $x_1 = p_1 u$ and $x_2 = p_2 u$.
Uniqueness: Given any $v : X \to U$, if $p_1 v = x_1$ and $p_2 v = x_2$, then $v = u$.

Proposition 2.17. *Products are unique up to isomorphism.*

Proof. Suppose

$$A \xleftarrow{\ p_1\ } P \xrightarrow{\ p_2\ } B$$

and

$$A \xleftarrow{\ q_1\ } Q \xrightarrow{\ q_2\ } B$$

are products of A and B. Then, since Q is a product, there is a unique $i : P \to Q$ such that $q_1 \circ i = p_1$ and $q_2 \circ i = p_2$. Similarly, since P is a product, there is a unique $j : Q \to P$ such that $p_1 \circ j = q_1$ and $p_2 \circ j = q_2$.

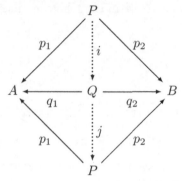

Composing, $p_1 \circ j \circ i = p_1$ and $p_2 \circ j \circ i = p_2$. Since also $p_1 \circ 1_P = p_1$ and $p_2 \circ 1_P = p_2$, it follows from the uniqueness condition that $j \circ i = 1_P$. Similarly, we can show $i \circ j = 1_Q$. Thus, $i : P \to Q$ is an isomorphism. $\quad\square$

If A and B have a product, we write

$$A \xleftarrow{\ p_1\ } A \times B \xrightarrow{\ p_2\ } B$$

for one such product. Then given X, x_1, x_2 as in the definition, we write

$$\langle x_1, x_2 \rangle \ \text{ for } \ u : X \to A \times B.$$

Note, however, that a pair of objects may have many different products in a category. For example, given a product $A \times B, p_1, p_2$, and any iso $h : A \times B \to Q$, the diagram $Q, p_1 \circ h, p_2 \circ h$ is also a product of A and B.

Now an arrow *into* a product

$$f : X \to A \times B$$

is "the same thing" as a pair of arrows

$$f_1 : X \to A, \qquad f_2 : X \to B.$$

So we can essentially forget about such arrows, in that they are uniquely determined by pairs of arrows. But something useful *is* gained if a category has products; namely, consider arrows *out of* the product,

$$g : A \times B \to Y.$$

Such a g is a "function in two variables"; given any two generalized elements $f_1 : X \to A$ and $f_2 : X \to B$, we have an element $g\langle f_1, f_2 \rangle : X \to Y$. Such arrows $g : A \times B \to Y$ are not "reducible" to anything more basic, the way arrows into products were (to be sure, they are related to the notion of an "exponential" Y^B, via "currying" $\lambda f : A \to Y^B$; we discuss this further in Chapter 6).

2.5   Examples of products

1. We have already seen cartesian products of sets. Note that if we choose a different definition of ordered pairs $\langle a, b \rangle$, we get different sets

   $$A \times B \quad \text{and} \quad A \times' B$$

   each of which is (part of) a product, and so are isomorphic. For instance, we could set

   $$\langle a, b \rangle = \{\{a\}, \{a, b\}\}$$
   $$\langle a, b \rangle' = \langle a, \langle a, b \rangle \rangle.$$

2. Products of "structured sets" like monoids or groups can often be constructed as products of the underlying sets with *componentwise* operations: If G and H are groups, for instance, $G \times H$ can be constructed by taking the underlying set of $G \times H$ to be the set $\{\langle g, h \rangle \mid g \in G, \ h \in H\}$ and defining the binary operation by

   $$\langle g, h \rangle \cdot \langle g', h' \rangle = \langle g \cdot g', h \cdot h' \rangle$$

   the unit by

   $$u = \langle u_G, u_H \rangle$$

   and inverses by

   $$\langle a, b \rangle^{-1} = \langle a^{-1}, b^{-1} \rangle.$$

   The projection homomorphisms $G \times H \to G$ (or H) are the evident ones $\langle g, h \rangle \mapsto g$ (or h).

3. Similarly, for categories $\mathbf{C}$ and $\mathbf{D}$, we already defined the category of pairs of objects and arrows,

$$\mathbf{C} \times \mathbf{D}.$$

Together with the evident projection functors, this is indeed a product in $\mathbf{Cat}$ (when $\mathbf{C}$ and $\mathbf{D}$ are small). (Check this: verify the UMP for the product category so defined.)

As special cases, we also get products of posets and of monoids as products of categories. (Check this: the projections and unique paired function are always monotone and so the product of posets, constructed in $\mathbf{Cat}$, is also a product in $\mathbf{Pos}$, and similarly for $\mathbf{Mon}$.)

4. Let P be a poset and consider a product of elements $p, q \in P$. We must have projections

$$p \times q \leq p$$
$$p \times q \leq q$$

and if for any element x,

$$x \leq p, \quad \text{and} \quad x \leq q$$

then we need

$$x \leq p \times q.$$

Do you recognize this operation $p \times q$? It is just what is usually called the *greatest lower bound*: $p \times q = p \wedge q$. Many other order-theoretic notions are also special cases of categorical ones, as we shall see later.

5. (For those who know something about Topology.) Let us show that the product of two *topological spaces* X, Y, as usually defined, really is a product in $\mathbf{Top}$, the category of spaces and continuous functions. Thus, suppose we have spaces X and Y and the product spaces $X \times Y$ with its projections

$$X \xleftarrow{p_1} X \times Y \xrightarrow{p_2} Y.$$

Recall that $O(X \times Y)$ is generated by basic open sets of the form $U \times V$ where $U \in O(X)$ and $V \in O(Y)$, so every $W \in O(X \times Y)$ is a union of such basic opens.

- Clearly p_1 is continuous, since $p_1^{-1}U = U \times Y$.
- Given any continuous $f_1 : Z \to X, f_2 : Z \to Y$, let $f : Z \to X \times Y$ be the function $f = \langle f_1, f_2 \rangle$. We just need to see that f is continuous.

- Given any $W = \bigcup_i (U_i \times V_i) \in O(X \times Y)$, $f^{-1}(W) = \bigcup_i f^{-1}(U_i \times V_i)$, so it suffices to show $f^{-1}(U \times V)$ is open. But

$$f^{-1}(U \times V) = f^{-1}((U \times Y) \cap (X \times V))$$
$$= f^{-1}(U \times Y) \cap f^{-1}(X \times V)$$
$$= f^{-1} \circ p_1^{-1}(U) \cap f^{-1} \circ p_2^{-1}(V)$$
$$= (f_1)^{-1}(U) \cap (f_2)^{-1}(V)$$

where $(f_1)^{-1}(U)$ and $(f_2)^{-1}(V)$ are open, since f_1 and f_2 are continuous.

The following diagram concisely captures the situation at hand:

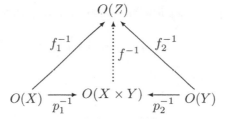

6. (For those familiar with type theory.) Let us consider the *category of types* of the (simply typed) λ-calculus. The λ–calculus is a formalism for the specification and manipulation of functions, based on the notions of "binding of variables" and functional evaluation. For example, given the real polynomial expression $x^2 + 2y$, in the λ-calculus one writes $\lambda y.x^2 + 2y$ for the function $y \mapsto x^2 + 2y$ (for each fixed value x), and $\lambda x \lambda y.x^2 + 2y$ for the function-valued function $x \mapsto (y \mapsto x^2 + 2y)$.

Formally, the λ-calculus consists of

- Types: $A \times B$, $A \to B, \ldots$ (generated from some basic types)
- Terms:

$$x, y, z, \ldots : A \quad \text{(variables for each type } A\text{)}$$
$$a : A, \ b : B, \ldots \quad \text{(possibly some typed constants)}$$
$$\langle a, b \rangle : A \times B \quad (a : A, \ b : B)$$
$$\text{fst}(c) : A \quad (c : A \times B)$$
$$\text{snd}(c) : B \quad (c : A \times B)$$
$$ca : B \quad (c : A \to B, \ a : A)$$
$$\lambda x.b : A \to B \quad (x : A, \ b : B)$$

- Equations:

$$\mathrm{fst}(\langle a,b\rangle) = a$$
$$\mathrm{snd}(\langle a,b\rangle) = b$$
$$\langle \mathrm{fst}(c),\mathrm{snd}(c)\rangle = c$$
$$(\lambda x.b)a = b[a/x]$$
$$\lambda x.cx = c \quad (\text{no } x \text{ in } c)$$

The relation $a \sim b$ (usually called $\beta\eta$-equivalence) on terms is defined to be the equivalence relation generated by the equations, and renaming of bound variables:

$$\lambda x.b = \lambda y.b[y/x] \quad (\text{no } y \text{ in } b)$$

The category of types $\mathbf{C}(\lambda)$ is now defined as follows:

- Objects: the types,
- Arrows $A \to B$: closed terms $c : A \to B$, identified if $c \sim c'$,
- Identities: $1_A = \lambda x.x$ (where $x : A$),
- Composition: $c \circ b = \lambda x.c(bx)$.

Let us verify that this is a well-defined category:
Unit laws:

$$c \circ 1_B = \lambda x(c((\lambda y.y)x)) = \lambda x(cx) = c$$
$$1_C \circ c = \lambda x((\lambda y.y)(cx)) = \lambda x(cx) = c$$

Associativity:

$$\begin{aligned}
c \circ (b \circ a) &= \lambda x(c((b \circ a)x)) \\
&= \lambda x(c((\lambda y.b(ay))x)) \\
&= \lambda x(c(b(ax))) \\
&= \lambda x(\lambda y(c(by))(ax)) \\
&= \lambda x((c \circ b)(ax)) \\
&= (c \circ b) \circ a
\end{aligned}$$

This category has binary products. Indeed, given types A and B, let

$$p_1 = \lambda z.\mathrm{fst}(z), \quad p_2 = \lambda z.\mathrm{snd}(z) \quad (z : A \times B).$$

And given a and b as in

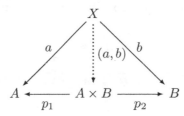

let

$$(a, b) = \lambda x.\langle ax, bx \rangle.$$

Then

$$\begin{aligned}
p_1 \circ (a, b) &= \lambda x(p_1((\lambda y.\langle ay, by\rangle)x)) \\
&= \lambda x(p_1\langle ax, bx\rangle) \\
&= \lambda x(ax) \\
&= a.
\end{aligned}$$

Similarly, $p_2 \circ (a, b) = b$.

Finally, if $c : X \to A \times B$ also has

$$p_1 \circ c = a, \qquad p_2 \circ c = b$$

then

$$\begin{aligned}
(a, b) &= \lambda x.\langle ax, bx \rangle \\
&= \lambda x.\langle (p_1 \circ c)x, (p_2 \circ c)x \rangle \\
&= \lambda x.\langle (\lambda y(p_1(cy)))x, (\lambda y(p_2(cy)))x \rangle \\
&= \lambda x.\langle (\lambda y((\lambda z.\mathrm{fst}(z))(cy)))x, (\lambda y((\lambda z.\mathrm{snd}(z))(cy)))x \rangle \\
&= \lambda x.\langle \lambda y(\mathrm{fst}(cy))x, \lambda y(\mathrm{snd}(cy))x \rangle \\
&= \lambda x.\langle \mathrm{fst}(cx), \mathrm{snd}(cx) \rangle \\
&= \lambda x.(cx) \\
&= c.
\end{aligned}$$

Remark 2.18. The λ-calculus had another surprising interpretation, namely as a system of notation for proofs in propositional calculus; this is known as the "Curry–Howard" correspondence. Briefly, the idea is that one interprets types as propositions (with $A \times B$ being conjunction and $A \to B$ implication) and terms $a : A$ as proofs of the proposition A. The term-forming rules such as

$$\frac{a : A \qquad b : B}{\langle a, b \rangle : A \times B}$$

can then be read as annotated rules of inference, showing how to build up labels for proofs inductively. So, for instance, a natural deduction proof such as

$$\frac{\dfrac{[A] \qquad [B]}{A \times B}}{\dfrac{B \to (A \times B)}{A \to (B \to (A \times B))}}$$

with square brackets indicating cancellation of premisses, is labeled as follows:

$$\frac{\dfrac{[x : A] \qquad [y : B]}{\langle x, y \rangle : A \times B}}{\dfrac{\lambda y. \langle x, y \rangle : B \to (A \times B)}{\lambda x \lambda y. \langle x, y \rangle : A \to (B \to (A \times B))}}$$

The final "proof term" $\lambda x \lambda y. \langle x, y \rangle$ thus records the given proof of the "proposition" $A \to (B \to (A \times B))$, and a different proof of the same proposition would give a different term.

Although one often speaks of a resulting "isomorphism" between logic and type theory, what we in fact have here is simply a functor from the category of proofs in the propositional calculus with conjunction and implication (as defined in example 10 of Section 1.4.), into the category of types of the λ-calculus. The functor will not generally be an isomorphism unless we impose some further equations between proofs.

2.6 Categories with products

Let **C** be a category that has a product diagram for every pair of objects. Suppose we have objects and arrows

$$
\begin{array}{ccccc}
A & \xleftarrow{\ p_1\ } & A \times A' & \xrightarrow{\ p_2\ } & A' \\
\downarrow{\scriptstyle f} & & & & \downarrow{\scriptstyle f'} \\
B & \xleftarrow{\ q_1\ } & B \times B' & \xrightarrow{\ q_2\ } & B'
\end{array}
$$

with indicated products. Then, we write

$$f \times f' : A \times A' \to B \times B'$$

for $f \times f' = \langle f \circ p_1, f' \circ p_2 \rangle$. Thus, both squares in the following diagram commute:

$$
\begin{array}{ccccc}
A & \xleftarrow{\;p_1\;} & A \times A' & \xrightarrow{\;p_2\;} & A' \\
\downarrow{\scriptstyle f} & & \vdots{\scriptstyle f \times f'} & & \downarrow{\scriptstyle f'} \\
B & \xleftarrow[\;q_1\;]{} & B \times B' & \xrightarrow[\;q_2\;]{} & B'
\end{array}
$$

In this way, if we choose a product for each pair of objects, we get a functor

$$\times : \mathbf{C} \times \mathbf{C} \to \mathbf{C}$$

as the reader can easily check, using the UMP of the product. A category which has a product for every pair of objects is said to *have binary products*.

We can also define ternary products

$$A_1 \times A_2 \times A_3$$

with an analogous UMP (there are three projections $p_i : A_1 \times A_2 \times A_3 \to A_i$, and for any object X and three arrows $x_i : X \to A_i$, there is a unique arrow $u : X \to A_1 \times A_2 \times A_3$ such that $p_i u = x_i$ for each of the three i's). Plainly, such a condition can be formulated for any number of factors.

It is clear, however, that if a category has binary products, then it has all finite products with two or more factors; for instance, one could set

$$A \times B \times C = (A \times B) \times C$$

to satisfy the UMP for ternary products. On the other hand, one could instead have taken $A \times (B \times C)$ just as well. This shows that the binary product operation $A \times B$ is associative up to isomorphism, for we must have

$$(A \times B) \times C \cong A \times (B \times C)$$

by the UMP of ternary products.

Observe also that a terminal object is a "nullary" product, that is, a product of no objects:

> Given no objects, there is an object 1 with no maps, and given any other object X and no maps, there is a unique arrow
>
> $$! : X \to 1$$
>
> making nothing further commute.

Similarly, any object A is the *unary product* of A with itself one time.

Finally, one can also define the product of a family of objects $(C_i)_{i \in I}$ indexed by *any* set I, by giving a UMP for "I-ary products" analogous to those for nullary, unary, binary, and n-ary products. We leave the precise formulation of this UMP as an exercise.

Definition 2.19. A category $\mathbf{C}$ is said to *have all finite products*, if it has a terminal object and all binary products (and therewith products of any finite cardinality). The category $\mathbf{C}$ *has all (small) products* if every set of objects in $\mathbf{C}$ has a product.

2.7   Hom-sets

In this section, we assume that all categories are locally small.

Recall that in any category $\mathbf{C}$, given any objects A and B, we write

$$\mathrm{Hom}(A, B) = \{f \in \mathbf{C} \mid f : A \to B\}$$

and call such a set of arrows a *Hom-set*. Note that any arrow $g : B \to B'$ in $\mathbf{C}$ induces a function

$$\mathrm{Hom}(A, g) : \mathrm{Hom}(A, B) \to \mathrm{Hom}(A, B')$$

$$(f : A \to B) \mapsto (g \circ f : A \to B \to B')$$

Thus, $\mathrm{Hom}(A, g) = g \circ f$; one sometimes writes g_* instead of $\mathrm{Hom}(A, g)$, so

$$g_*(f) = g \circ f.$$

Let us show that this determines a functor

$$\mathrm{Hom}(A, -) : \mathbf{C} \to \mathbf{Sets},$$

called the (covariant) *representable functor* of A. We need to show that

$$\mathrm{Hom}(A, 1_X) = 1_{\mathrm{Hom}(A,X)}$$

and that

$$\mathrm{Hom}(A, g \circ f) = \mathrm{Hom}(A, g) \circ \mathrm{Hom}(A, f).$$

Taking an argument $x : A \to X$, we clearly have

$$\mathrm{Hom}(A, 1_X)(x) = 1_X \circ x$$

$$= x$$

$$= 1_{\mathrm{Hom}(A,X)}(x)$$

and

$$\mathrm{Hom}(A, g \circ f)(x) = (g \circ f) \circ x$$

$$= g \circ (f \circ x)$$

$$= \mathrm{Hom}(A, g)(\mathrm{Hom}(A, f)(x)).$$

We will study such representable functors much more carefully later. For now, we just want to see how one can use Hom-sets to give another formulation of the definition of products.

For any object P, a pair of arrows $p_1 : P \to A$ and $p_2 : P \to B$ determine an element (p_1, p_2) of the set

$$\mathrm{Hom}(P, A) \times \mathrm{Hom}(P, B).$$

Now, given any arrow

$$x : X \to P$$

composing with p_1 and p_2 gives a pair of arrows $x_1 = p_1 \circ x : X \to A$ and $x_2 = p_2 \circ x : X \to B$, as indicated in the following diagram:

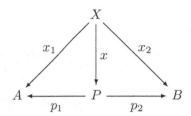

In this way, we have a function

$$\vartheta_X = (\mathrm{Hom}(X, p_1), \mathrm{Hom}(X, p_2)) : \mathrm{Hom}(X, P) \to \mathrm{Hom}(X, A) \times \mathrm{Hom}(X, B)$$

defined by

$$\vartheta_X(x) = (x_1, x_2) \tag{2.1}$$

This function ϑ_X can be used to express concisely the condition of being a product as follows.

Proposition 2.20. *A diagram of the form*

$$A \xleftarrow{\quad p_1 \quad} P \xrightarrow{\quad p_2 \quad} B$$

is a product for A and B iff for every object X, the canonical function ϑ_X given in (2.1) is an isomorphism,

$$\vartheta_X : \mathrm{Hom}(X, P) \cong \mathrm{Hom}(X, A) \times \mathrm{Hom}(X, B).$$

Proof. Examine the UMP of the product: it says exactly that for every element $(x_1, x_2) \in \mathrm{Hom}(X, A) \times \mathrm{Hom}(X, B)$, there is a unique $x \in \mathrm{Hom}(X, P)$ such that $\vartheta_X(x) = (x_1, x_2)$, that is, ϑ_X is bijective. $\qquad\square$

Definition 2.21. Let $\mathbf{C}$, $\mathbf{D}$ be categories with binary products. A functor $F : \mathbf{C} \to \mathbf{D}$ is said to *preserve binary products* if it takes every product diagram

$$A \xleftarrow{\quad p_1 \quad} A \times B \xrightarrow{\quad p_2 \quad} B \qquad\qquad \text{in } \mathbf{C}$$

to a product diagram

$$FA \xleftarrow{\quad Fp_1 \quad} F(A \times B) \xrightarrow{\quad Fp_2 \quad} FB \qquad\qquad \text{in } \mathbf{D}.$$

It follows that F preserves products just if

$$F(A \times B) \cong FA \times FB$$

"canonically," that is, iff the canonical "comparison arrow"

$$\langle Fp_1, Fp_2 \rangle : F(A \times B) \to FA \times FB$$

in $\mathbf{D}$ is an iso.

For example, the forgetful functor $U : \mathbf{Mon} \to \mathbf{Sets}$ preserves binary products.

Corollary 2.22. *For any object X in a category $\mathbf{C}$ with products, the (covariant) representable functor*

$$\mathrm{Hom}_{\mathbf{C}}(X, -) : \mathbf{C} \to \mathbf{Sets}$$

preserves products.

Proof. For any $A, B \in \mathbf{C}$, the foregoing proposition 2.20 says that there is a canonical isomorphism:

$$\mathrm{Hom}_{\mathbf{C}}(X, A \times B) \cong \mathrm{Hom}_{\mathbf{C}}(X, A) \times \mathrm{Hom}_{\mathbf{C}}(X, B)$$

$\square$

2.8 Exercises

1. Show that a function between sets is an epimorphism if and only if it is surjective. Conclude that the isos in **Sets** are exactly the epi-monos.

2. Show that in a poset category, all arrows are both monic and epic.

3. (Inverses are unique.) If an arrow $f : A \to B$ has inverses $g, g' : B \to A$ (i. e., $g \circ f = 1_A$ and $f \circ g = 1_B$, and similarly for g'), then $g = g'$.

4. With regard to a commutative triangle,

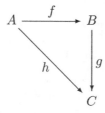

in any category $\mathbf{C}$, show

(a) if f and g are isos (resp. monos, resp. epis), so is h;

(b) if h is monic, so is f;

(c) if h is epic, so is g;

(d) (by example) if h is monic, g need not be.

5. Show that the following are equivalent for an arrow

$$f : A \to B$$

in any category:

  (a) f is an isomorphism.

  (b) f is both a mono and a split epi.

  (c) f is both a split mono and an epi.

  (d) f is both a split mono and a split epi.

6. Show that a homomorphism $h : G \to H$ of graphs is monic just if it is injective on both edges and vertices.

7. Show that in any category, any retract of a projective object is also projective.

8. Show that all sets are projective (use the axiom of choice).

9. Show that the epis among posets are the surjections (on elements), and that the one-element poset **1** is projective.

10. Show that sets, regarded as discrete posets, are projective in the category of posets (use the foregoing exercises). Give an example of a poset that is not projective. Show that every projective poset is discrete, that is, a set. Conclude that **Sets** is (isomorphic to) the "full subcategory" of projectives in **Pos**, consisting of all projective posets and all monotone maps between them.

11. Let A be a set. Define an *A-monoid* to be a monoid M equipped with a function $m : A \to U(M)$ (to the underlying set of M). A morphism $h : (M, m) \to (N, n)$ of A-monoids is to be a monoid homomorphism $h : M \to N$ such that $U(h) \circ m = n$ (a commutative triangle). Together with the evident identities and composites, this defines a category A-**Mon** of A-monoids.

    Show that an initial object in A-**Mon** is the same thing as a free monoid $M(A)$ on A. (Hint: compare their respective UMPs.)

12. Show that for any Boolean algebra B, Boolean homomorphisms $h : B \to \mathbf{2}$ correspond exactly to ultrafilters in B.

13. In any category with binary products, show directly that

$$A \times (B \times C) \cong (A \times B) \times C.$$

14. (a) For any index set I, define the product $\prod_{i \in I} X_i$ of an I-indexed family of objects $(X_i)_{i \in I}$ in a category, by giving a UMP generalizing that for binary products (the case $I = 2$).

   (b) Show that in **Sets**, for any set X the set X^I of all functions $f : I \to X$ has this UMP, with respect to the "constant family" where $X_i = X$

for all $i \in I$, and thus

$$X^I \cong \prod_{i \in I} X.$$

15. Given a category $\mathbf{C}$ with objects A and B, define the category $\mathbf{C}_{A,B}$ to have objects (X, x_1, x_2), where $x_1 : X \to A$, $x_2 : X \to B$, and with arrows $f : (X, x_1, x_2) \to (Y, y_1, y_2)$ being arrows $f : X \to Y$ with $y_1 \circ f = x_1$ and $y_2 \circ f = x_2$.

Show that $\mathbf{C}_{A,B}$ has a terminal object just in case A and B have a product in $\mathbf{C}$.

16. In the category of types $\mathbf{C}(\lambda)$ of the λ-calculus, determine the product functor $A, B \mapsto A \times B$ explicitly. Also show that, for any fixed type A, there is a functor $A \to (-) : \mathbf{C}(\lambda) \to \mathbf{C}(\lambda)$, taking any type X to $A \to X$.

17. In any category $\mathbf{C}$ with products, define the *graph* of an arrow $f : A \to B$ to be the monomorphism

$$\Gamma(f) = \langle 1_A, f \rangle : A \rightarrowtail A \times B$$

(Why is this monic?). Show that for $\mathbf{C} = \mathbf{Sets}$ this determines a functor $\Gamma : \mathbf{Sets} \to \mathbf{Rel}$ to the category $\mathbf{Rel}$ of relations, as defined in the exercises to Chapter 1. (To get an actual relation $R(f) \subseteq A \times B$, take the image of $\Gamma(f) : A \rightarrowtail A \times B$.)

18. Show that the forgetful functor $U : \mathbf{Mon} \to \mathbf{Sets}$ from monoids to sets is representable. Infer that U preserves all (small) products.

3

DUALITY

We have seen a few examples of definitions and statements that exhibit a kind of "duality," like initial and terminal object and epimorphisms and monomorphisms. We now want to consider this duality more systematically. Despite its rather trivial first impression, it is indeed a deep and powerful aspect of the categorical approach to mathematical structures.

3.1 The duality principle

First, let us look again at the formal definition of a category: There are two kinds of things, objects A, B, C and ..., arrows $f, g, h, \ldots$; four operations $\mathrm{dom}(f)$, $\mathrm{cod}(f)$, 1_A, $g \circ f$; and these satisfy the following seven axioms:

$$\mathrm{dom}(1_A) = A \qquad \mathrm{cod}(1_A) = A$$
$$f \circ 1_{\mathrm{dom}(f)} = f \qquad 1_{\mathrm{cod}(f)} \circ f = f \tag{3.1}$$
$$\mathrm{dom}(g \circ f) = \mathrm{dom}(f) \qquad \mathrm{cod}(g \circ f) = \mathrm{cod}(g)$$
$$h \circ (g \circ f) = (h \circ g) \circ f$$

The operation "$g \circ f$" is only defined where

$$\mathrm{dom}(g) = \mathrm{cod}(f),$$

so a suitable form of this should occur as a condition on each equation containing $\circ$, as in $\mathrm{dom}(g) = \mathrm{cod}(f) \Rightarrow \mathrm{dom}(g \circ f) = \mathrm{dom}(f)$.

Now, given any sentence Σ in the elementary language of category theory, we can form the "dual statement" Σ^* by making the following replacements:

$$f \circ g \quad \text{for} \quad g \circ f$$
$$\mathrm{cod} \quad \text{for} \quad \mathrm{dom}$$
$$\mathrm{dom} \quad \text{for} \quad \mathrm{cod}.$$

It is easy to see that then Σ^* will also be a well-formed sentence. Next, suppose we have shown a sentence Σ to entail one Δ, that is, $\Sigma \Rightarrow \Delta$, without using any of the category axioms, then clearly $\Sigma^* \Rightarrow \Delta^*$, since the substituted terms are treated as mere undefined constants. But now observe that the axioms (3.1) for category theory (CT) are themselves "self-dual," in the sense that we have,

$$CT^* = CT.$$

We therefore have the following *duality principle.*

Proposition 3.1 (formal duality). *For any sentence* Σ *in the language of category theory, if* Σ *follows from the axioms for categories, then so does its dual* Σ^*:

$$CT \Rightarrow \Sigma \quad \text{implies} \quad CT \Rightarrow \Sigma^*$$

Taking a more conceptual point of view, note that if a statement Σ involves some diagram of objects and arrows,

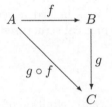

then the dual statement Σ^* involves the diagram obtained from it by reversing the direction and the order of compositions of arrows.

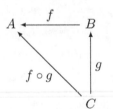

Recalling the opposite category $\mathbf{C}^{\mathrm{op}}$ of a category $\mathbf{C}$, we see that an interpretation of a statement Σ in $\mathbf{C}$ automatically gives an interpretation of Σ^* in $\mathbf{C}^{\mathrm{op}}$.

Now suppose that a statement Σ holds for all categories $\mathbf{C}$. Then it also holds in all categories $\mathbf{C}^{\mathrm{op}}$, and so Σ^* holds in all categories $(\mathbf{C}^{\mathrm{op}})^{\mathrm{op}}$. But since for every category $\mathbf{C}$,

$$(\mathbf{C}^{\mathrm{op}})^{\mathrm{op}} = \mathbf{C}, \tag{3.2}$$

we see that Σ^* also holds in all categories $\mathbf{C}$. We therefore have the following conceptual form of the duality principle.

Proposition 3.2 (Conceptual duality). *For any statement* Σ *about categories, if* Σ *holds for all categories, then so does the dual statement* Σ^*.

It may seem that only very simple or trivial statements, such as "terminal objects are unique up to isomorphism" are going to be subject to this sort of duality, but in fact this is far from being so. Categorical duality turns out to be a very powerful and a far-reaching phenomenon, as we shall see. Like the duality between points and lines in projective geometry, it effectively doubles ones "bang for the buck," yielding two theorems for every proof.

One way this occurs is that, rather than considering statements about all categories, we can also consider the dual of an abstract definition of a structure or property of objects and arrows, like "being a product diagram." The dual structure or property is arrived at by reversing the order of composition and the words "domain" and "codomain." (Equivalently, it results from interpreting the original property in the opposite category.) Section 3.2 provides an example of this kind.

3.2 Coproducts

Let us consider the example of products and see what the dual notion must be. First, recall the definition of a product.

Definition 3.3. A diagram $A \xleftarrow{p_1} P \xrightarrow{p_2} B$ is a *product* of A and B, if for any Z and $A \xleftarrow{z_1} Z \xrightarrow{z_2} B$ there is a unique $u : Z \to P$ with $p_i \circ u = z_i$, all as indicated in

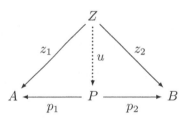

Now what is the dual statement?

A diagram $A \xrightarrow{q_1} Q \xleftarrow{q_2} B$ is a "dual-product" of A and B if for any Z and $A \xrightarrow{z_1} Z \xleftarrow{z_2} B$ there is a unique $u : Q \to Z$ with $u \circ q_i = z_i$, all as indicated in

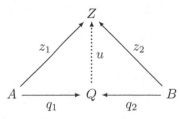

Actually, these are called *coproducts*; the convention is to use the prefix "co-" to indicate the dual notion. We usually write $A \xrightarrow{i_1} A+B \xleftarrow{i_2} B$ for the coproduct and $[f, g]$ for the uniquely determined arrow $u : A + B \to Z$. The "coprojections" $i_1 : A \to A + B$ and $i_2 : B \to A + B$ are usually called *injections*, even though they need not be "injective" in any sense.

A coproduct of two objects is therefore exactly their product in the opposite category. Of course, this immediately gives lots of examples of coproducts. But what about some more familiar ones?

Example 3.4. In **Sets**, the coproduct $A + B$ of two sets is their disjoint union, which can be constructed, for example, as

$$A + B = \{(a, 1) \mid a \in A\} \cup \{(b, 2) \mid b \in B\}$$

with evident coproduct injections

$$i_1(a) = (a, 1), \qquad i_2(b) = (b, 2).$$

Given any functions f and g as in

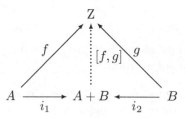

we define

$$[f, g](x, \delta) = \begin{cases} f(x) & \delta = 1 \\ g(x) & \delta = 2. \end{cases}$$

Then, if we have an h with $h \circ i_1 = f$ and $h \circ i_2 = g$, then for any $(x, \delta) \in A + B$, we must have

$$h(x, \delta) = [f, g](x, \delta)$$

as can be easily calculated.

Note that in **Sets**, every finite set A is a coproduct:

$$A \cong 1 + 1 + \cdots + 1 \quad (n\text{-times})$$

for $n = \operatorname{card}(A)$. This is because a function $f : A \to Z$ is uniquely determined by its values $f(a)$ for all $a \in A$. So we have

$$A \cong \{a_1\} + \{a_n\} + \cdots + \{a_n\}$$
$$\cong 1 + 1 + \cdots + 1 \quad (n\text{-times}).$$

In this spirit, we often write simply $2 = 1 + 1, 3 = 1 + 1 + 1$, etc.

Example 3.5. If $M(A)$ and $M(B)$ are *free* monoids on sets A and B, then in **Mon** we can construct their coproduct as

$$M(A) + M(B) \cong M(A + B).$$

One can see this directly by considering words over $A + B$, but it also follows abstractly by using the diagram

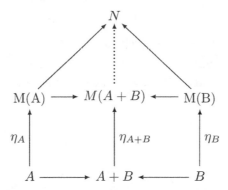

in which the η's are the respective insertions of generators. The universal mapping properties (UMPs) of $M(A)$, $M(B)$, $A+B$, and $M(A+B)$ then imply that the last of these has the required UMP of $M(A)+M(B)$. Note that the set of elements of the coproduct $M(A) + M(B)$ of $M(A)$ and $M(B)$ is *not* the coproduct of the underlying sets, but is only *generated by* the coproduct of their generators, $A + B$. We shall consider coproducts of arbitrary, that is, not necessarily free, monoids presently.

The foregoing example says that the free monoid functor $M :$ **Sets** $\rightarrow$ **Mon** preserves coproducts. This is an instance of a much more general phenomenon, which we consider later, related to the fact we have already seen that the forgetful functor $U :$ **Mon** $\rightarrow$ **Sets** is representable and so preserves products.

Example 3.6. In **Top**, the coproduct of two spaces

$$X + Y$$

is their disjoint union with the topology $O(X + Y) \cong O(X) \times O(Y)$. Note that this follows the pattern of discrete spaces, for which $O(X) = P(X) \cong 2^X$. Thus, for discrete spaces, we indeed have

$$O(X + Y) \cong 2^{X+Y} \cong 2^X \times 2^Y \cong O(X) \times O(Y).$$

A related fact is that the product of two powerset Boolean algebras $\mathcal{P}(A)$ and $\mathcal{P}(B)$ is also a powerset, namely of the coproduct of the sets A and B,

$$\mathcal{P}(A) \times \mathcal{P}(B) \cong \mathcal{P}(A + B).$$

We leave the verification as an exercise.

Coproducts of posets are similarly constructed from the coproducts of the underlying sets, by "putting them side by side." What about "rooted" posets, that is, posets with a distinguished initial element 0? In the category Pos$_0$ of such posets and monotone maps that preserve 0, one constructs the coproduct of two such posets A and B from the coproduct $A + B$ in the category Pos of posets, by "identifying" the two different 0s,

$$A +_{\text{Pos}_0} B = (A +_{\text{Pos}} B)/\text{"}0_A = 0_B\text{"}.$$

We shall soon see how to describe such identifications (quotients of equivalence relations) as "coequalizers."

Example 3.7. In a fixed poset P, what is a coproduct of two elements $p, q \in P$? We have

$$p \leq p + q \quad \text{and} \quad q \leq p + q$$

and if

$$p \leq z \quad \text{and} \quad q \leq z$$

then

$$p + q \leq z.$$

So $p + q = p \vee q$ is the *join*, or "least upper bound," of p and q.

Example 3.8. In the *category of proofs* of a deductive system of logic of example 10, Section 1.4, the usual natural deduction rules of disjunction introduction and elimination give rise to coproducts. Specifically, the introduction rules,

$$\frac{\varphi}{\varphi \vee \psi} \qquad \frac{\psi}{\varphi \vee \psi}$$

determine arrows $i_1 : \varphi \to \varphi \vee \psi$ and $i_2 : \psi \to \varphi \vee \psi$, and the elimination rule,

$$\frac{\varphi \vee \psi \qquad \overset{[\varphi]}{\overset{\vdots}{\vartheta}} \qquad \overset{[\psi]}{\overset{\vdots}{\vartheta}}}{\vartheta}$$

turns a pair of arrows $p : \varphi \to \vartheta$ and $q : \psi \to \vartheta$ into an arrow $[p, q] : \varphi \vee \psi \to \vartheta$. The required equations,

$$[p, q] \circ i_1 = p \qquad [p, q] \circ i_2 = q \tag{3.3}$$

will evidently not hold, however, since we are taking identity of proofs as identity of arrows. In order to get coproducts, then, we need to "force" these equations to hold by passing to equivalence classes of proofs, under the equivalence relation generated by these equations, together with the complementary one,

$$[r \circ i_1, r \circ i_2] = r \tag{3.4}$$

for any $r : \varphi \vee \psi \to \vartheta$. (The intuition behind these identifications is that one should equate proofs which become the same when one omits such "detours.")

In the new category with equivalence classes of proofs as arrows, the arrow $[p, q]$ will also be the *unique* one satisfying (3.3), so that $\varphi \vee \psi$ indeed becomes a coproduct.

Closely related to this example (via the Curry–Howard correspondence of remark 2.18) are the sum types in the λ-calculus, as usually formulated using **case** terms; these are coproducts in the *category of types* defined in Section 2.5.

Example 3.9. Two monoids A, B have a coproduct of the form

$$A + B = M(|A| + |B|)/\sim$$

where, as before, the free monoid $M(|A|+|B|)$ is strings (words) over the disjoint union $|A| + |B|$ of the underlying sets—that is, the elements of A and B—and the equivalence relation $v \sim w$ is the least one containing all instances of the following equations:

$$(\ldots x\, u_A\, y \ldots) = (\ldots x\, y \ldots)$$
$$(\ldots x\, u_B\, y \ldots) = (\ldots x\, y \ldots)$$
$$(\ldots a\, a' \ldots) = (\ldots a \cdot_A a' \ldots)$$
$$(\ldots b\, b' \ldots) = (\ldots b \cdot_B b' \ldots).$$

(If you need a refresher on quotienting a set by an equivalence relation, skip ahead and read the beginning of Section 3.4 now.) The unit is, of course, the equivalence class $[-]$ of the empty word (which is the same as $[u_A]$ and $[u_B]$). Multiplication of equivalence classes is also as expected, namely

$$[x \ldots y] \cdot [x' \ldots y'] = [x \ldots yx' \ldots y'].$$

The coproduct injections $i_A : A \to A + B$ and $i_B : B \to A + B$ are simply

$$i_A(a) = [a], \qquad i_B(b) = [b],$$

which are now easily seen to be homomorphisms. Given any homomorphisms $f : A \to M$ and $g : B \to M$ into a monoid M, the unique homomorphism

$$[f, g] : A + B \longrightarrow M$$

is defined by first extending the function $[|f|, |g|] : |A| + |B| \to |M|$ to one $[f, g]'$ on the free monoid $M(|A| + |B|)$,

$$|A| + |B| \xrightarrow{\;[|f|, |g|]\;} |M|$$

$$M(|A| + |B|) \xrightarrow{\;[f, g]'\;} M$$
$$\downarrow \qquad \qquad \nearrow {[f, g]}$$
$$M(|A| + |B|)/\sim$$

and then observing that $[f, g]'$ "respects the equivalence relation $\sim$," in the sense that if $v \sim w$ in $M(|A| + |B|)$, then $[f, g]'(v) = [f, g]'(w)$. Thus, the map $[f, g]'$ extends to the quotient to yield the desired map $[f, g] : M(|A| + |B|)/\sim \longrightarrow M$. (Why is this homomorphism the *unique* one $h : M(|A| + |B|)/\sim \longrightarrow M$ with $hi_A = f$ and $hi_B = g$?) Summarizing, we thus have

$$A + B \cong M(|A| + |B|)/\sim .$$

This construction also works to give coproducts in **Groups**, where it is usually called the *free product* of A and B and written $A \oplus B$, as well as many other categories of "algebras," that is, sets equipped with operations. Again, as in the free case, the underlying set of $A + B$ is *not* the coproduct of A and B as sets (the forgetful functor **Mon** → **Sets** does not preserve coproducts).

Example 3.10. For *abelian groups* A, B, the free product $A \oplus B$ need not be abelian. One could, of course, take a further quotient of $A \oplus B$ to get a coproduct in the category Ab of abelian groups, but there is a more convenient (and important) presentation, which we now consider.

Since the words in the free product $A \oplus B$ must be forced to satisfy the further commutativity conditions

$$(a_1 b_1 b_2 a_2 \ldots) \sim (a_1 a_2 \ldots b_1 b_2 \ldots)$$

we can shuffle all the a's to the front, and the b's to the back, of the words. But, furthermore, we already have

$$(a_1 a_2 \ldots b_1 b_2 \ldots) \sim (a_1 + a_2 + \cdots + b_1 + b_2 + \cdots).$$

Thus, we in effect have pairs of elements (a, b). So we can take the *product* set as the underlying set of the coproduct

$$|A + B| = |A \times B|.$$

As inclusions, we use the homomorphisms

$$i_A(a) = (a, 0_B)$$
$$i_B(b) = (0_A, b).$$

Then, given any homomorphisms $A \xrightarrow{f} X \xleftarrow{g} B$, we let $[f, g] : A + B \to X$ be defined by

$$[f, g](a, b) = f(a) +_X g(b)$$

which can easily be seen to do the trick (exercise!).

Moreover, not only can the underlying *sets* be the same, the product and coproduct of abelian groups are actually isomorphic as *groups*.

Proposition 3.11. *In the category* **Ab** *of abelian groups, there is a canonical isomorphism between the binary coproduct and product,*

$$A + B \cong A \times B.$$

Proof. To define an arrow $\vartheta : A + B \to A \times B$, we need one $A \to A \times B$ (and one $B \to A \times B$), so we need arrows $A \to A$ and $A \to B$ (and $B \to A$ and $B \to B$). For these, we take $1_A : A \to A$ and the zero homomorphism $0_B : A \to B$ (and $0_A : B \to A$ and $1_B : B \to B$). Thus, all together, we get

$$\vartheta = [\langle 1_A, 0_B \rangle, \langle 0_A, 1_B \rangle] : A + B \to A \times B.$$

Then given any $(a, b) \in A + B$, we have

$$\begin{aligned}
\vartheta(a, b) &= [\langle 1_A, 0_B \rangle, \langle 0_A, 1_B \rangle](a, b) \\
&= \langle 1_A, 0_B \rangle(a) + \langle 0_A, 1_B \rangle(b) \\
&= (1_A(a), 0_B(a)) + (0_A(b), 1_B(b)) \\
&= (a, 0_B) + (0_A, b) \\
&= (a + 0_A, 0_B + b) \\
&= (a, b).
\end{aligned}$$

$\square$

This fact was first observed by Mac Lane, and it was shown to lead to a binary operation of addition on parallel arrows $f, g : A \to B$ between abelian groups (and related structures like modules and vector spaces). In fact, the group structure of a particular abelian group A can be recovered from this operation on arrows into A. More generally, the existence of such an addition operation on arrows can be used as the basis of an abstract description of categories like **Ab**, called "abelian categories," which are suitable for axiomatic homology theory.

Just as with products, one can consider the empty coproduct, which is an initial object 0, as well as coproducts of several factors, and the coproduct of two arrows,

$$f + f' : A + A' \to B + B'$$

which leads to a coproduct functor $+ : \mathbf{C} \times \mathbf{C} \to \mathbf{C}$ on categories $\mathbf{C}$ with binary coproducts. All of these facts follow simply by duality; that is, by considering the dual notions in the opposite category. Similarly, we have the following proposition.

Proposition 3.12. *Coproducts are unique up to isomorphism.*

Proof. Use duality and the fact that the dual of "isomorphism" is "isomorphism."

$\square$

In just the same way, one also shows that binary coproducts are associative up to isomorphism, $(A + B) + C \cong A + (B + C)$.

Thus is general, in the future it will suffice to introduce new notions once and then simply observe that the dual notions have analogous (but dual) properties. Sections 3.3 and 3.4 give another example of this sort.

3.3 Equalizers

In this section, we consider another abstract characterization; this time a common generalization of the kernel of a homomorphism and an equationally defined "variety," like the set of zeros of a real-valued function—as well as set theory's axiom of separation.

Definition 3.13. In any category **C**, given parallel arrows

$$A \underset{g}{\overset{f}{\rightrightarrows}} B$$

an *equalizer* of f and g consists of an object E and an arrow $e : E \to A$, universal such that

$$f \circ e = g \circ e.$$

That is, given any $z : Z \to A$ with $f \circ z = g \circ z$, there is a *unique* $u : Z \to E$ with $e \circ u = z$, all as in the following diagram.

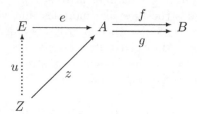

Let us consider some simple examples.

Example 3.14. Suppose we have the functions $f, g : \mathbb{R}^2 \rightrightarrows \mathbb{R}$, where

$$f(x,y) = x^2 + y^2$$
$$g(x,y) = 1$$

and we take the equalizer, say in **Top**. This is the subspace,

$$S = \{(x,y) \in \mathbb{R}^2 \mid x^2 + y^2 = 1\} \hookrightarrow \mathbb{R}^2,$$

that is, the unit circle in the plane. For, given any "generalized element" $z : Z \to \mathbb{R}^2$, we get a pair of such "elements" $z_1, z_2 : Z \to \mathbb{R}$ just by composing with the two projections, $z = \langle z_1, z_2 \rangle$, and for these we then have

$$f(z) = g(z) \text{ iff } z_1{}^2 + z_2{}^2 = 1$$
$$\text{iff } "\langle z_1, z_2 \rangle = z \in S",$$

where the last line really means that there is a factorization $z = \bar{z} \circ i$ of z through the inclusion $i : S \hookrightarrow \mathbb{R}^2$, as indicated in the following diagram:

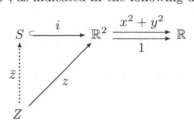

Since the inclusion i is monic, such a factorization, if it exists, is necessarily unique, and thus $S \hookrightarrow \mathbb{R}^2$ is indeed the equalizer of f and g.

Example 3.15. Similarly, in **Sets**, given any functions $f, g : A \rightrightarrows B$, their equalizer is the inclusion into A of the equationally defined subset

$$\{x \in A \mid f(x) = g(x)\} \hookrightarrow A.$$

The argument is essentially the same as the one just given.

Let us pause here to note that in fact, every subset $U \subseteq A$ is of this "equatio-nal" form, that is, every subset is an equalizer for some pair of functions. Indeed, one can do this in a very canonical way. First, let us put

$$2 = \{\top, \bot\},$$

thinking of it as the set of "truth values." Then consider the *characteristic function*

$$\chi_U : A \to 2,$$

defined for $x \in A$ by

$$\chi_U(x) = \begin{cases} \top & x \in U \\ \bot & x \notin U. \end{cases}$$

Thus, we have

$$U = \{x \in A \mid \chi_U(x) = \top\}.$$

So the following is an equalizer:

$$U \longrightarrow A \underset{\chi_U}{\overset{\top!}{\rightrightarrows}} 2$$

where $\top! = \top \circ ! : U \xrightarrow{!} 1 \xrightarrow{\top} 2$.

Moreover, for every function,

$$\varphi : A \to 2$$

we can form the "variety" (i.e., equational subset)

$$V_\varphi = \{x \in A \mid \varphi(x) = \top\}$$

as an equalizer, in the same way. (Thinking of φ as a "propositional function" defined on A, the subset $V_\varphi \subseteq A$ is the "extension" of φ provided by the axiom of separation.)

Now, it is easy to see that these operations χ_U and V_φ are mutually inverse:

$$V_{\chi_U} = \{x \in A \mid \chi_U(x) = \top\}$$
$$= \{x \in A \mid x \in U\}$$
$$= U$$

for any $U \subseteq A$, and given any $\varphi : A \to 2$,

$$\chi_{V_\varphi}(x) = \begin{cases} \top & x \in V_\varphi \\ \bot & x \notin V_\varphi \end{cases}$$
$$= \begin{cases} \top & \varphi(x) = \top \\ \bot & \varphi(x) = \bot \end{cases}$$
$$= \varphi(x).$$

Thus, we have the familiar isomorphism

$$\mathrm{Hom}(A, 2) \cong P(A),$$

mediated by taking equalizers.

The fact that equalizers of functions can be taken to be subsets is a special case of a more general phenomenon.

Proposition 3.16. *In any category, if $e : E \to A$ is an equalizer of some pair of arrows, then e is monic.*

Proof. Consider the diagram

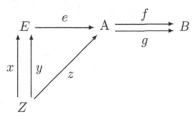

in which we assume e is the equalizer of f and g. Supposing $ex = ey$, we want to show $x = y$. Put $z = ex = ey$. Then $fz = fex = gex = gz$, so there is a *unique* $u : Z \to E$ such that $eu = z$. So from $ex = z$ and $ey = z$ it follows that $x = u = y$. $\square$

Example 3.17. In many other categories, such as posets and monoids, the equalizer of a parallel pair of arrows $f, g : A \rightrightarrows B$ can be constructed by taking the equalizer of the underlying functions as above, that is, the subset $A(f = g) \subseteq A$ of elements $x \in A$ where f and g agree, $f(x) = g(x)$, and then restricting the structure of A to $A(f = g)$. For instance, in posets one takes the ordering from A restricted to this subset $A(f = g)$, and in topological spaces one takes the subspace topology.

In monoids, the subset $A(f = g)$ is then also a monoid with the operations from A, and the inclusion is therefore a homomorphism. This is so because $f(u_A) = u_B = g(u_A)$, and if $f(a) = g(a)$ and $f(a') = g(a')$, then $f(a \cdot a') = f(a) \cdot f(a') = g(a) \cdot g(a') = g(a \cdot a')$. Thus, $A(f = g)$ contains the unit and is closed under the product operation.

In abelian groups, for instance, one has an alternate description of the equalizer, using the fact that,

$$f(x) = g(x) \quad \text{iff} \quad (f - g)(x) = 0.$$

Thus, the equalizer of f and g is the same as that of the homomorphism $(f - g)$ and the zero homomorphism $0 : A \to B$, so it suffices to consider equalizers of the special form $A(h, 0) \rightarrowtail A$ for arbitrary homomorphisms $h : A \to B$. This subgroup of A is called the *kernel* of h, written $\ker(h)$. Thus, we have the equalizer

$$\ker(f - g) \lhook\joinrel\longrightarrow A \underset{g}{\overset{f}{\rightrightarrows}} B.$$

The kernel of a homomorphism is of fundamental importance in the study of groups, as we consider further in Chapter 4.

3.4 Coequalizers

A coequalizer is a generalization of a quotient by an equivalence relation, so let us begin by reviewing that notion, which we have already made use of several times. Recall first that an *equivalence relation* on a set X is a binary relation $x \sim y$, which is

reflexive: $x \sim x$,
symmetric: $x \sim y$ implies $y \sim x$,
transitive: $x \sim y$ and $y \sim z$ implies $x \sim z$.

Given such a relation, define the *equivalence class* $[x]$ of an element $x \in X$ by

$$[x] = \{y \in X \mid x \sim y\}.$$

The various different equivalence classes $[x]$ then form a *partition* of X, in the sense that every element y is in exactly one of them, namely $[y]$ (prove this!).

One sometimes thinks of an equivalence relation as arising from the equivalent elements having some property in common (like being the same color). One can then regard the equivalence classes $[x]$ as the properties and in that sense as "abstract objects" (the colors red, blue, etc., themselves). This is sometimes known as "definition by abstraction," and it describes, for example, the way that the real numbers can be constructed from Cauchy sequences of rationals or the finite cardinal numbers from finite sets.

The set of all equivalence classes

$$X/\sim \; = \; \{[x] \mid x \in X\}$$

may be called the *quotient* of X by $\sim$. It is used in place of X when one wants to "abstract away" the difference between equivalent elements $x \sim y$, in the sense that in $X/\sim$ such elements (and only such) are identified, since

$$[x] = [y] \quad \text{iff} \quad x \sim y.$$

Observe that the *quotient mapping*,

$$q : X \longrightarrow X/\sim$$

taking x to $[x]$ has the property that a map $f : X \to Y$ extends along q,

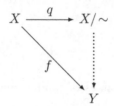

just in case f respects the equivalence relation, in the sense that $x \sim y$ implies $f(x) = f(y)$.

Now let us consider the notion dual to that of an equalizer, namely that of a coequalizer.

Definition 3.18. For any parallel arrows $f, g : A \to B$ in a category $\mathbf{C}$, a *coequalizer* consists of Q and $q : B \to Q$, universal with the property $qf = qg$, as in

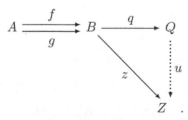

That is, given any Z and $z : B \to Z$, if $zf = zg$, then there exists a unique $u : Q \to Z$ such that $uq = z$.

First, observe that by duality, we know that such a coequalizer q in a category $\mathbf{C}$ is an equalizer in $\mathbf{C}^{\mathrm{op}}$, hence monic by proposition 3.16, and so epic in $\mathbf{C}$.

Proposition 3.19. *If $q : B \to Q$ is a coequalizer of some pair of arrows, then q is epic.*

We can therefore think of a coequalizer $q : B \twoheadrightarrow Q$ as a "collapse" of B by "identifying" all pairs $f(a) = g(a)$ (speaking as if there were such "elements" $a \in A$). Moreover, we do this in the "minimal" way, that is, disturbing B as little as possible, in that one can always map Q to anything else Z in which all such identifications hold.

Example 3.20. Let $R \subseteq X \times X$ be an equivalence relation on a set X, and consider the diagram

$$R \underset{r_2}{\overset{r_1}{\rightrightarrows}} X$$

where the r's are the two projections of the inclusion $R \subseteq X \times X$,

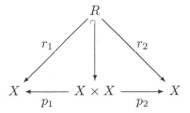

The quotient projection

$$\pi : X \longrightarrow X/R$$

defined by $x \mapsto [x]$ is then a coequalizer of r_1 and r_2. For given an $f : X \to Y$ as in

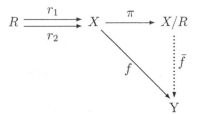

there exists a function $\bar{f}$ such that

$$\bar{f}\pi(x) = f(x)$$

whenever f respects R in the sense that $(x, x') \in R$ implies $f(x) = f(x')$, as already noted. But this condition just says that $f \circ r_1 = f \circ r_2$, since $f \circ r_1(x, x') = f(x)$ and $f \circ r_2(x, x') = f(x')$ for all $(x, x') \in R$. Moreover, if it exists, such a function $\bar{f}$, is then necessarily unique, since π is an epimorphism.

The coequalizer in **Sets** of an arbitrary parallel pair of functions $f \cdot g : A \rightrightarrows B$ can be constructed by quotienting B by the equivalence relation generated by the equations $f(x) = g(x)$ for all $x \in A$. We leave the details as an exercise.

Example 3.21. In example 3.6, we considered the coproduct of *rooted* posets P and Q by first making $P + Q$ in posets and then "identifying" the resulting two different 0-elements 0_P and 0_Q (i.e., the images of these under the respective coproduct inclusions. We can now describe this "identification" as a coequalizer, taken in posets,

$$1 \begin{array}{c} \xrightarrow{\;0_P\;} \\ \xrightarrow[\;0_Q\;]{} \end{array} P + Q \longrightarrow P + Q/(0_P = 0_Q).$$

This clearly has the right UMP to be the coproduct in *rooted* posets.

In topology one also often makes "identifications" of points (as in making the circle out of the interval by identifying the endpoints), of subspaces (making the torus from a region in the plane, etc.). These and many similar "gluing" constructions can be described as coequalizers. In **Top**, the coequalizer of a parallel pair of maps $f, g : X \to Y$ can be constructed as a *quotient space* of Y (see the exercises).

Example 3.22. *Presentations of algebras*
Consider any category of "algebras," that is, sets equipped with operations (of finite arity), such as monoids or groups. We shall show later that such a category has free algebras for all sets and coequalizers for all parallel pairs of arrows (see the exercises for a proof that monoids have coequalizers). We can use these to determine the notion of a *presentation* of an algebra by *generators* and *relations*. For example, suppose we are given

$$\text{Generators: } x, y, z$$

$$\text{Relations: } xy = z, \ y^2 = 1. \tag{3.5}$$

To build an algebra on these generators and satisfying these relations, start with the free algebra,

$$F(3) = F(x, y, z),$$

and then "force" the relation $xy = z$ to hold by taking a coequalizer of the maps

$$F(1) \begin{array}{c} \xrightarrow{\;xy\;} \\ \xrightarrow[\;z\;]{} \end{array} F(3) \xrightarrow{\;\;q\;\;} Q.$$

We use the fact that maps $F(1) \to A$ correspond to elements $a \in A$ by $v \mapsto a$, where v is the single generator of $F(1)$. Now similarly, for the equation $y^2 = 1$, take the coequalizer

$$F(1) \begin{array}{c} \xrightarrow{\;q(y^2)\;} \\ \xrightarrow[\;q(1)\;]{} \end{array} Q \longrightarrow Q'.$$

These two steps can actually be done simultaneously. Let

$$F(2) = F(1) + F(1)$$

$$F(2) \underset{g}{\overset{f}{\rightrightarrows}} F(3)$$

where $f = [xy, y^2]$ and $g = [z, 1]$. The coequalizer $q : F(3) \to Q$ of f and g then "forces" both equations to hold, in the sense that in Q, we have

$$q(x)q(y) = q(z), \quad q(y)^2 = 1.$$

Moreover, no other relations among the generators hold in Q except those required to hold by the stipulated equations. To make the last statement precise, observe that given any algebra A and any three elements $a, b, c \in A$ such that $ab = c$ and $b^2 = 1$, by the UMP of Q there is a unique homomorphism $u : Q \to A$ such that

$$u(x) = a, \qquad u(y) = b, \qquad u(z) = c.$$

Thus, any other equation that holds among the generators in Q will also hold in any other algebra in which the stipulated equations (3.5) hold, since the homomorphism u also preserves equations. In this sense, Q is the "universal" algebra with three generators satisfying the stipulated equations; as may be written suggestively in the form

$$Q \cong F(x, y, z)/(xy = z, \ y^2 = 1).$$

Generally, given a finite presentation

$$\text{Generators:} \quad g_1, \dots, g_n$$
$$\text{Relations:} \quad l_1 = r_1, \ \dots, \ l_m = r_m \qquad\qquad (3.6)$$

(where the l_i and r_i are arbitrary terms built from the generators and the operations) the algebra determined by that presentation is the coequalizer

$$F(m) \underset{r}{\overset{l}{\rightrightarrows}} F(n) \longrightarrow Q = F(n)/(l = r)$$

where $l = [l_1, \dots, l_m]$ and $r = [r_1, \dots, r_m]$. Moreover, any such coequalizer between (finite) free algebras can clearly be regarded as a (finite) presentation by generators and relations. Algebras that can be given in this way are said to be *finitely presented.*

Warning 3.23. Presentations are not unique. One may well have two different presentations $F(n)/(l = r)$ and $F(n')/(l' = r')$ by generators and relations of the same algebra,

$$F(n)/(l = r) \cong F(n')/(l' = r').$$

For instance, given $F(n)/(l = r)$ just add a new generator g_{n+1} and the new relation $g_n = g_{n+1}$. In general, there are many different ways of presenting a given algebra, just like there are many ways of axiomatizing a logical theory.

We did not really make use of the finiteness condition in the foregoing considerations. Indeed, *any* sets of generators G and relations R give rise to an algebra
in the same way, by taking the coequalizer

$$F(R) \underset{r_2}{\overset{r_1}{\rightrightarrows}} F(G) \longrightarrow F(G)/(r_1 = r_2).$$

In fact, every algebra can be "presented" by generators and relations in this
sense, that is, as a coequalizer of maps between free algebras. Specifically, we
have the following proposition for monoids, an analogous version of which also
holds for groups and other algebras.

Proposition 3.24. *For every monoid M there are sets R and G and a
coequalizer diagram,*

$$F(R) \underset{r_2}{\overset{r_1}{\rightrightarrows}} F(G) \longrightarrow M$$

with $F(R)$ and $F(G)$ free; thus, $M \cong F(G)/(r_1 = r_2)$.

Proof. For any monoid N, let us write $TN = M(|N|)$ for the free monoid on
the set of elements of N (and note that T is therefore a functor). There is a
homomorphism,

$$\pi : TN \to N$$

$$\pi(x_1, \ldots, x_n) = x_1 \cdot \ldots \cdot x_n$$

induced by the identity $1_{|N|} : |N| \to |N|$ on the generators. (Here we are writing
the elements of TN as tuples $(x_1, \ldots, x_n)$ rather than strings $x_1 \ldots x_n$ for clarity.)
Applying this construction twice to a monoid M results in the arrows π and ε
in the following diagram:

$$T^2 M \underset{\mu}{\overset{\varepsilon}{\rightrightarrows}} TM \xrightarrow{\ \pi\ } M \qquad\qquad (3.7)$$

where $T^2 M = TTM$ and $\mu = T\pi$. Explicitly, the elements of $T^2 M$ are tuples of tuples of elements of M, say $((x_1, \ldots, x_n), \ldots, (z_1, \ldots, z_m))$, and the
homomorphisms ε and μ have the effect:

$$\varepsilon((x_1, \ldots, x_n), \ldots, (z_1, \ldots, z_m)) = (x_1, \ldots, x_n, \ldots, z_1, \ldots, z_m)$$

$$\mu((x_1, \ldots, x_n), \ldots, (z_1, \ldots, z_m)) = (x_1 \cdot \ldots \cdot x_n, \ldots, z_1 \cdot \ldots \cdot z_m)$$

Briefly, ε uses the multiplication in TM and μ uses that in M.

Now clearly $\pi \circ \varepsilon = \pi \circ \mu$. We claim that (3.7) is a coequalizer of monoids.
To that end, suppose we have a monoid N and a homomorphism $h : TM \to N$

with $h\varepsilon = h\mu$. Then for any tuple $(x, \ldots, z)$, we have

$$h(x, \ldots, z) = h\varepsilon((x, \ldots, z))$$
$$= h\mu((x, \ldots, z)) \qquad (3.8)$$
$$= h(x \cdot \ldots \cdot z).$$

Now define $\bar{h} = h \circ i$, where $i : |M| \to |TM|$ is the insertion of generators, as indicated in the following:

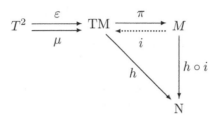

We then have

$$\bar{h}\pi(x, \ldots, z) = hi\pi(x, \ldots, z)$$
$$= h(x \cdot \ldots \cdot z)$$
$$= h(x, \ldots, z) \qquad \text{by (3.8)}.$$

We leave it as an easy exercise for the reader to show that $\bar{h}$ is a homomorphism.

□

3.5 Exercises

1. In any category $\mathbf{C}$, show that

$$A \xrightarrow{c_1} C \xleftarrow{c_2} B$$

is a coproduct diagram just if for every object Z, the map

$$\mathrm{Hom}(C, Z) \longrightarrow \mathrm{Hom}(A, Z) \times \mathrm{Hom}(B, Z)$$
$$f \longmapsto \quad \langle f \circ c_1, \ f \circ c_2 \rangle$$

is an isomorphism. Do this by using duality, taking the corresponding fact about products as given.

2. Show in detail that the free monoid functor M preserves coproducts: for any sets A, B,

$$M(A) + M(B) \cong M(A + B) \qquad \text{(canonically)}.$$

Do this as indicated in the text by using the UMPs of the coproducts $A + B$ and $M(A) + M(B)$ and of free monoids.

3. Verify that the construction given in the text of the coproduct of monoids $A + B$ as a quotient of the free monoid $M(|A| + |B|)$ really is a coproduct in the category of monoids.

4. Show that the product of two powerset Boolean algebras $\mathcal{P}(A)$ and $\mathcal{P}(B)$ is also a powerset, namely of the coproduct of the sets A and B,

$$\mathcal{P}(A) \times \mathcal{P}(B) \cong \mathcal{P}(A + B).$$

(Hint: determine the projections $\pi_1 : \mathcal{P}(A + B) \to \mathcal{P}(A)$ and $\pi_2 : \mathcal{P}(A + B) \to \mathcal{P}(B)$, and check that they have the UMP of the product.)

5. Consider the category of proofs of a natural deduction system with disjunction introduction and elimination rules. Identify proofs under the equations

$$[p, q] \circ i_1 = p, \qquad [p, q] \circ i_2 = q$$
$$[r \circ i_1, r \circ i_2] = r$$

for any $p : A \to C$, $q : B \to C$, and $r : A + B \to C$. By passing to equivalence classes of proofs with respect to the equivalence relation generated by these equations (i.e., two proofs are equivalent if you can get one from the other by removing all such "detours"). Show that the resulting category does indeed have coproducts.

6. Verify that the category of monoids has all equalizers and finite products, then do the same for abelian groups.

7. Show that in any category with coproducts, the coproduct of two projectives is again projective.

8. Dualize the notion of projectivity to define an *injective* object in a category. Show that a map of posets is monic iff it is injective on elements. Give examples of a poset that is injective and one that is not injective.

9. Complete the proof of proposition 3.24 in the text by showing that $\bar{h}$ is indeed a homomorphism.

10. In the proof of proposition 3.24 in the text it is shown that any monoid M has a specific presentation $T^2M \rightrightarrows TM \to M$ as a coequalizer of free monoids. Show that coequalizers of this particular form are preserved by the forgetful functor $\mathbf{Mon} \to \mathbf{Sets}$.

11. Prove that $\mathbf{Sets}$ has all coequalizers by constructing the coequalizer of a parallel pair of functions,

$$A \underset{g}{\overset{f}{\rightrightarrows}} B \longrightarrow Q = B/(f = g)$$

by quotienting B by a suitable equivalence relation R on B, generated by the pairs $(f(x), g(x))$ for all $x \in A$. (Define R to be the intersection of all equivalence relations on B containing all such pairs.)

12. Verify the coproduct–coequalizer construction mentioned in the text for coproducts of rooted posets, that is, posets with a least element 0 and monotone maps preserving 0. Specifically, show that the coproduct $P +_0 Q$ of two such posets can be constructed as a coequalizer in posets,

$$1 \underset{0_Q}{\overset{0_P}{\rightrightarrows}} P + Q \longrightarrow P +_0 Q.$$

(You may assume as given the fact that the category of posets has all coequalizers.)

13. Show that the category of monoids has all coequalizers as follows.
1. Given any pair of monoid homomorphisms $f, g : M \to N$, show that the following equivalence relations on N agree:

(a) $n \sim n' \Leftrightarrow$ for all monoids X and homomorphisms $h : N \to X$, one has $hf = hg$ implies $hn = hn'$,

(b) the intersection of all equivalence relations $\sim$ on N satisfying $fm \sim gm$ for all $m \in M$ as well as

$$n \sim n' \text{ and } m \sim m' \Rightarrow n \cdot m \sim n' \cdot m'$$

2. Taking $\sim$ to be the equivalence relation defined in (1), show that the quotient set $N/\sim$ is a monoid under $[n] \cdot [m] = [n \cdot m]$, and the projection $N \to N/\sim$ is the coequalizer of f and g.

14. Consider the following category of sets:

(a) Given a function $f : A \to B$, describe the equalizer of the functions $f \circ p_1, f \circ p_2 : A \times A \to B$ as a (binary) relation on A and show that it is an equivalence relation (called the *kernel* of f).

(b) Show that the kernel of the quotient $A \to A/R$ by an equivalence relation R is R itself.

(c) Given *any* binary relation $R \subseteq A \times A$, let $\langle R \rangle$ be the equivalence relation on A generated by R (the least equivalence relation on A containing R). Show that the quotient $A \to A/\langle R \rangle$ is the coequalizer of the two projections $R \rightrightarrows A$.

(d) Using the foregoing, show that for any binary relation R on a set A, one can characterize the equivalence relation $\langle R \rangle$ generated by R as the kernel of the coequalizer of the two projections of R.

15. Construct coequalizers in **Top** as follows. Given a parallel pair of maps $f, g : X \rightrightarrows Y$, make a *quotient space* $q : Y \to Q$ by (i) taking the coequalizer of $|f|$ and $|g|$ in **Sets** to get the function $|q| : |Y| \to |Q|$, then (ii) equip $|Q|$ with the *quotient topology*, under which a set $V \subseteq Q$ is open iff $q^{-1}(V) \subseteq Y$ is open. This is plainly the finest topology on $|Q|$ that makes the projection $|q|$ continuous.

4

GROUPS AND CATEGORIES

This chapter is devoted to some of the various connections between groups and categories. If you already know the basic group theory covered here, then this gives you some insight into the categorical constructions we have learned so far; and if you do not know it yet, then you learn it now as an application of category theory. We focus on three different aspects of the relationship between categories and groups:

1. groups in a category,
2. the category of groups,
3. groups as categories.

4.1 Groups in a category

As we have already seen, the notion of a group arises as an abstraction of the automorphisms of an object. In a specific, concrete case, a group G may thus consist of certain arrows $g : X \to X$ for some object X in a category $\mathbf{C}$,

$$G \subseteq \mathrm{Hom}_{\mathbf{C}}(X, X)$$

But the abstract group concept can also be described directly as an object in a category, equipped with a certain structure. This more subtle notion of a "group in a category" also proves to be quite useful.

Let $\mathbf{C}$ be a category with finite products. The notion of a group in $\mathbf{C}$ essentially generalizes the usual notion of a group in **Sets**.

Definition 4.1. A *group* in $\mathbf{C}$ consists of objects and arrows as so:

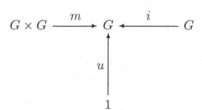

satisfying the following conditions:

1. m is associative, that is, the following commutes:

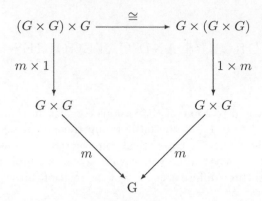

where $\cong$ is the canonical associativity isomorphism for products.

2. u is a unit for m, that is, both triangles in the following commute:

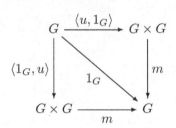

where we write u for the "constant arrow" $u! : G \xrightarrow{!} 1 \xrightarrow{u} G$.

3. i is an inverse with respect to m, that is, both sides of the following commute:

$$
\begin{array}{ccccc}
G \times G & \xleftarrow{\;\Delta\;} & G & \xrightarrow{\;\Delta\;} & G \times G \\
{\scriptstyle 1_G \times i}\downarrow & & \downarrow{\scriptstyle u} & & \downarrow{\scriptstyle i \times 1_G} \\
G \times G & \xrightarrow{\;m\;} & G & \xleftarrow{\;m\;} & G \times G
\end{array}
$$

where $\Delta = \langle 1_G, 1_G \rangle$.

Note that the requirement that these diagrams commute is equivalent to the more familiar condition that, for all (generalized) elements,

$$x, y, z : \; Z \to G$$

the following equations hold:

$$m(m(x,y),z) = m(x, m(y,z))$$
$$m(x,u) = x = m(u,x)$$
$$m(x,ix) = u = m(ix,x)$$

Definition 4.2. A *homomorphism* $h : G \to H$ of groups in **C** consists of an arrow in **C**,

$$h : G \to H$$

such that

1. h preserves m:

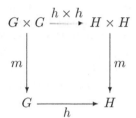

2. h preserves u:

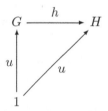

3. h preserves i:

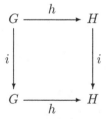

With the evident identities and composites, we thus have a category of groups in **C**, denoted by

$$\text{Group}(\mathbf{C})$$

Example 4.3. The idea of an internal group in a category captures the familiar notion of a group with additional structure.

- A group in the usual sense is a group in the category **Sets**.
- A topological group is a group in **Top**, the category of topological spaces.
- A (partially) ordered group is a group in the category **Pos** of posets (in this case, the inverse operation is usually required to be order-*reversing*, that is, of the form $i : G^{op} \to G$).

For example, the real numbers $\mathbb{R}$ under addition are a topological and an ordered group, since the operations of addition $x + y$ and additive inverse $-x$ are continuous and order-preserving (resp. reversing). They are a topological "semigroup" under multiplication $x \cdot y$ as well, but the multiplicative inverse operation $1/x$ is not continuous (or even defined!) at 0.

Example 4.4. Suppose we have a group G in the category **Groups** of groups. So G is a group equipped with group homomorphisms $m : G \times G \to G$, etc. as in definition 4.1. Let us take this apart in more elementary terms. Write the multiplication of the group G, that is, on the underlying set $|G|$, as $x \circ y$ and write the homomorphic multiplication m as $x \star y$. That the latter *is* a homomorphism from the product group $G \times G$ to G says in particular that, for all $g, h \in G \times G$ we have $m(g \circ h) = m(g) \circ m(h)$. Recalling that $g = (g_1, g_2)$, $h = (h_1, h_2)$ and multiplication $\circ$ on $G \times G$ is pointwise, this then comes to the following:

$$(g_1 \circ h_1) \star (g_2 \circ h_2) = (g_1 \star g_2) \circ (h_1 \star h_2) \tag{4.1}$$

Write 1° for the unit with respect to $\circ$ and $1^\star$ for the unit of $\star$. The following proposition is called the "Eckmann–Hilton argument," and was first used in homotopy theory.

Proposition 4.5. *Given any set G equipped with two binary operations $\circ, \star :$ $G \times G \to G$ with units 1° and $1^\star$, respectively and satisfying (4.1), the following hold.*

1. $1^\circ = 1^\star$.

2. $\circ = \star$.

3. *The operation $\circ = \star$ is commutative.*

Proof. First, we have

$$1^\circ = 1^\circ \circ 1^\circ$$
$$= (1^\circ \star 1^\star) \circ (1^\star \star 1^\circ)$$
$$= (1^\circ \circ 1^\star) \star (1^\star \circ 1^\circ)$$
$$= 1^\star \star 1^\star$$
$$= 1^\star.$$

Thus, let us write $1^\circ = 1 = 1^\star$. Next, we have,

$$x \circ y = (x \star 1) \circ (1 \star y) = (x \circ 1) \star (1 \circ y) = x \star y.$$

Thus, let us write $x \circ y = x \cdot y = x \star y$. Finally, we have

$$x \cdot y = (1 \cdot x) \cdot (y \cdot 1) = (1 \cdot y) \cdot (x \cdot 1) = y \cdot x.$$

$\square$

We therefore have the following.

Corollary 4.6. *The groups in the category of groups are exactly the abelian groups.*

Proof. We have just shown that a group in **Groups** is necessarily abelian, so it just remains to see that any abelian group admits homomorphic group operations. We leave this as an easy exercise. $\square$

Remark 4.7. Note that we did not really need the full group structure in this argument. Indeed, the same result holds for monoids in the category of monoids: these are exactly the commutative monoids.

Example 4.8. A further example of an internal algebraic structure in a category is provided by the notion of a (strict) *monoidal category*.

Definition 4.9. A *strict monoidal category* is a category **C** equipped with a binary operation $\otimes : \mathbf{C} \times \mathbf{C} \to \mathbf{C}$ which is functorial and associative,

$$A \otimes (B \otimes C) = (A \otimes B) \otimes C, \qquad (4.2)$$

together with a distinguished object I that acts as a unit,

$$I \otimes C = C = C \otimes I. \qquad (4.3)$$

A strict monoidal category is exactly the same thing as a monoid in **Cat**. Examples where the underlying category is a poset P include both the meet $x \wedge y$ and join $x \vee y$ operations, with terminal object 1 and initial object 0 as units, respectively (assuming P has these structures), as well as the poset $\mathrm{End}(P)$ of monotone maps $f : P \to P$, ordered pointwise, with composition $g \circ f$ as $\otimes$ and 1_P as unit. A discrete monoidal category, that is, one with a discrete underlying category, is obviously just a regular monoid (in **Sets**), while a monoidal category with only one object is a monoidal monoid, and thus exactly a commutative monoid, by the foregoing remark 4.7.

More general strict monoidal categories, that is, ones having a proper category with many objects and arrows, are rather less common—not for a paucity of such structures, but because the required equations (4.2) and (4.3) typically hold only "up to isomorphism." This is so, for example, for products $A \times B$ and coproducts $A + B$, as well as many other operations like tensor products $A \otimes B$ of vector spaces, modules, algebras over a ring, etc. (the category of proofs in linear logic provides more examples). We return to this more general notion of a (not necessarily strict) monoidal category once we have the required notion of a "natural isomorphism" (in Chapter 7), which is required to make the above notion of "up to isomorphism" precise.

A basic example of a non-poset monoidal category that *is* strict is provided by the category $\mathbf{Ord}_{\mathrm{fin}}$ of all finite ordinal numbers $0, 1, 2, \ldots$, which can be represented in set theory as,

$$0 = \emptyset,$$
$$n + 1 = \{0, \ldots, n\}.$$

The arrows are just all functions between these sets. The monoidal product $m \otimes n$ is then $m+n$ and 0 is the unit. In a sense that can be made precise in the expected way, this example is in fact the "free monoidal category on one object."

In logical terms, the concept of an internal group corresponds to the observation that one can "model the theory of groups" in *any* category with finite products, not just **Sets**. Thus, for instance, one can also define the notion of a *group in the λ-calculus*, since the category of types of the λ-calculus also has finite products. Of course the same is true for other algebraic theories, like monoids and rings, given by operations and equations. Theories involving other logical operations like negations, implication, or quantifiers can be modeled in categories having more structure than just finite products. Here we have a glimpse of so-called *categorical semantics*. Such semantics can be useful for theories that are not complete with respect to models in **Sets**, such as certain theories in intuitionistic logic.

4.2 The category of groups

Let G and H be groups (in **Sets**), and let

$$h : G \to H$$

be a group homomorphism. The *kernel* of h is defined by the equalizer

$$\ker(h) = \{g \in G \mid h(g) = u\} \longrightarrow G \underset{u}{\overset{h}{\rightrightarrows}} H$$

where, again, we write $u : G \to H$ for the constant homomorphism

$$u! = G \overset{!}{\to} 1 \overset{u}{\to} H.$$

We have already seen that this specification makes the above an equalizer diagram.

Observe that $\ker(h)$ is a *subgroup*. Indeed, it is a *normal subgroup*, in the sense that for any $k \in \ker(h)$, we have (using multiplicative notation)

$$g \cdot k \cdot g^{-1} \in \ker(h) \quad \text{for all } g \in G.$$

Now if $N \overset{i}{\rightarrowtail} G$ is *any* normal subgroup, we can construct the *coequalizer*

$$N \underset{u}{\overset{i}{\rightrightarrows}} G \overset{\pi}{\longrightarrow} G/N$$

sending $g \in G$ to u iff $g \in N$ ("killing off N"), as follows: the elements of G/N are the "cosets of N," that is, equivalence classes of the form $[g]$ for all $g \in G$, where we define

$$g \sim h \quad \text{iff} \quad g \cdot h^{-1} \in N.$$

(Prove that this is an equivalence relation!) The multiplication on the *factor group* G/N is then given by

$$[x] \cdot [y] = [x \cdot y]$$

which is well defined since N is normal: given any u, v with $x \sim u$ and $y \sim v$, we have

$$x \cdot y \sim u \cdot v \iff (x \cdot y) \cdot (u \cdot v)^{-1} \in N$$

but

$$(x \cdot y) \cdot (u \cdot v)^{-1} = x \cdot y \cdot v^{-1} \cdot u^{-1}$$
$$= x \cdot (u^{-1} \cdot u) \cdot y \cdot v^{-1} \cdot u^{-1}$$
$$= (x \cdot u^{-1}) \cdot (u \cdot (y \cdot v^{-1}) \cdot u^{-1}),$$

the last of which is evidently in N.

Let us show that the diagram above really is a coequalizer. First, it is clear that

$$\pi \circ i = \pi \circ u!$$

since $n \cdot u = n$ implies $[n] = [u]$. Suppose we have $f : G \to H$ killing N, that is, $f(n) = u$ for all $n \in N$. We then propose a "factorization" $\bar{f}$, as indicated in

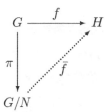

to be defined by

$$\bar{f}[g] = f(g).$$

This is well defined if $x \sim y$ implies $f(x) = f(y)$. But, since $x \sim y$ implies $f(x \cdot y^{-1}) = u$, we have

$$f(x) = f(x \cdot y^{-1} \cdot y) = f(x \cdot y^{-1}) \cdot f(y) = u \cdot f(y) = f(y).$$

Moreover, $\bar{f}$ is unique with $\pi\bar{f} = f$, since π is epic. Thus, we have shown most of the following classical *Homomorphism Theorem for Groups*.

Theorem 4.10. *Every group homomorphism $h : G \to H$ has a kernel $\ker(h) = h^{-1}(u)$, which is a normal subgroup of G with the property that, for any normal subgroup $N \subseteq G$*

$$N \subseteq \ker(h)$$

iff there is a (necessarily unique) homomorphism $\bar{h} : G/N \to H$ with $\bar{h} \circ \pi = h$, as indicated in the following diagram.

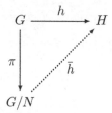

Proof. It only remains to show that if such a factorization $\bar{h}$ exists, then $N \subseteq \ker(h)$. But this is clear, since $\pi(N) = \{[u_G]\}$. So, $h(n) = \bar{h}\pi(n) = \bar{h}([n]) = u_H$. □

Finally, putting $N = \ker(h)$ in the theorem, and taking any $[x], [y] \in G/\ker(h)$, we have

$$\bar{h}[x] = \bar{h}[y] \;\Rightarrow\; h(x) = h(y)$$
$$\Rightarrow\; h(xy^{-1}) = u$$
$$\Rightarrow\; xy^{-1} \in \ker(h)$$
$$\Rightarrow\; x \sim y$$
$$\Rightarrow\; [x] = [y].$$

Thus, $\bar{h}$ is injective, and we conclude.

Corollary 4.11. *Every group homomorphism $h : G \to H$ factors as a quotient followed by an injective homomorphism,*

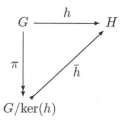

Thus, $\bar{h} : G/\ker(h) \xrightarrow{\sim} \operatorname{im}(h) \subseteq H$ is an isomorphism onto the subgroup $\operatorname{im}(h)$ that is the image of h.

In particular, therefore, a homomorphism h is injective if and only if its kernel is "trivial," in the sense that $\ker(h) = \{u\}$.

There is a dual to the notion of a kernel of a homomorphism $h : G \to H$, namely a cokernel $c : H \to C$, which is the universal way of "killing off h" in the sense that $c \circ h = u$. Cokernels are special coequalizers, in just the way that kernels are special equalizers. We leave the details as an exercise.

4.3 Groups as categories

First, let us recall that a group is a category. In particular, a group is a category with one object, in which every arrow is an iso. If G and H are groups, regarded as categories, then we can consider arbitrary functors between them

$$f : G \to H.$$

It is obvious that a functor between groups is exactly the same thing as a group homomorphism.

What is a functor $R : G \to \mathbf{C}$ from a group G to another category $\mathbf{C}$ that is not necessarily a group? If $\mathbf{C}$ is the category of (finite-dimensional) vector spaces and linear transformations, then such a functor is just what the group theorist calls a "linear representation" of G; such a representation permits the description of the group elements as matrices, and the group operation as matrix multiplication. In general, any functor $R : G \to \mathbf{C}$ may be regarded as a representation of G in the category $\mathbf{C}$: the elements of G become automorphisms of some object in $\mathbf{C}$. A permutation representation, for instance, is simply a functor into $\mathbf{Sets}$.

We now want to generalize the notions of kernel of a homomorphism, and quotient or factor group by a normal subgroup, from groups to arbitrary categories, and then give the analogous homomorphism theorem for categories.

Definition 4.12. A *congruence* on a category $\mathbf{C}$ is an equivalence relation $f \sim g$ on arrows such that

1. $f \sim g$ implies $\mathrm{dom}(f) = \mathrm{dom}(g)$ and $\mathrm{cod}(f) = \mathrm{cod}(g)$,

2. $f \sim g$ implies $bfa \sim bga$ for all arrows $a : A \to X$ and $b : Y \to B$, where $\mathrm{dom}(f) = X = \mathrm{dom}(g)$ and $\mathrm{cod}(f) = Y = \mathrm{cod}(g)$,

Let $\sim$ be a congruence on the category $\mathbf{C}$, and define the *congruence category* $\mathbf{C}^{\sim}$ by

$$(\mathbf{C}^{\sim})_0 = \mathbf{C}_0$$
$$(\mathbf{C}^{\sim})_1 = \{\langle f, g \rangle \mid f \sim g\}$$
$$\tilde{1}_C = \langle 1_C, 1_C \rangle$$
$$\langle f', g' \rangle \circ \langle f, g \rangle = \langle f'f, g'g \rangle$$

One easily checks that this composition is well defined, using the congruence conditions.

There are two evident projection functors:

$$\mathbf{C}^{\sim} \underset{p_2}{\overset{p_1}{\rightrightarrows}} \mathbf{C}$$

We build the *quotient category* $\mathbf{C}/\sim$ as follows:

$$(\mathbf{C}/\sim)_0 = \mathbf{C}_0$$
$$(\mathbf{C}/\sim)_1 = (\mathbf{C}_1)/\sim$$

The arrows have the form $[f]$ where $f \in \mathbf{C}_1$, and we can put $1_{[C]} = [1_C]$, and $[g] \circ [f] = [g \circ f]$, as is easily checked, again using the congruence conditions.

There is an evident quotient functor $\pi : \mathbf{C} \to \mathbf{C}/\sim$, making the following a coequalizer of categories:

$$\mathbf{C}^{\sim} \underset{p_2}{\overset{p_1}{\rightrightarrows}} \mathbf{C} \overset{\pi}{\longrightarrow} \mathbf{C}/\sim$$

This is proved much as for groups.

An exercise shows how to use this construction to make coequalizers for certain functors. Let us show how to use it to prove an analogous "homomorphism theorem for categories." Suppose we have categories $\mathbf{C}$ and $\mathbf{D}$ and a functor

$$F : \mathbf{C} \to \mathbf{D}.$$

Then, F determines a congruence $\sim_F$ on $\mathbf{C}$ by setting

$$f \sim_F g \quad \text{iff} \quad \text{dom}(f) = \text{dom}(g), \ \text{cod}(f) = \text{cod}(g), \ F(f) = F(g)$$

That this *is* a congruence is easily checked.

Let us write

$$\ker(F) = \mathbf{C}^{\sim_F} \rightrightarrows \mathbf{C}$$

for the congruence category, and call this the *kernel category* of F.

The quotient category

$$\mathbf{C}/\sim_F$$

then has the following universal mapping property (UMP):

Theorem 4.13. *Every functor $F : \mathbf{C} \to \mathbf{D}$ has a kernel category $\ker(F)$, determined by a congruence $\sim_F$ on $\mathbf{C}$ such that given any congruence $\sim$ on $\mathbf{C}$ one has*

$$f \sim g \Rightarrow f \sim_F g$$

if and only if there is a factorization $\widetilde{F} : \mathbf{C}/\sim \longrightarrow \mathbf{D}$, as indicated in

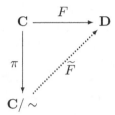

Just as in the case of groups, applying the theorem to the case $\mathbf{C}^\sim = \ker(F)$ gives a factorization theorem.

Corollary 4.14. *Every functor $F : \mathbf{C} \to \mathbf{D}$ factors as $F = \widetilde{F} \circ \pi$,*

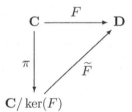

where π is bijective on objects and surjective on Hom-sets, and $\widetilde{F}$ is injective on Hom-sets (i.e., "faithful"):

$$\widetilde{F}_{A,B} : \mathrm{Hom}(A, B) \rightarrowtail \mathrm{Hom}(FA, FB) \quad \textit{for all } A, B \in \mathbf{C}/\ker(F)$$

4.4 Finitely presented categories

Finally, let us consider categories presented by generators and relations.

We begin with the free category $\mathbf{C}(G)$ on some finite graph G, and then consider a finite set Σ of relations of the form

$$(g_1 \circ \ldots \circ g_n) = (g'_1 \circ \ldots \circ g'_m)$$

with all $g_i \in G$, and $\mathrm{dom}(g_n) = \mathrm{dom}(g'_m)$ and $\mathrm{cod}(g_1) = \mathrm{cod}(g'_1)$. Such a relation identifies two "paths" in $\mathbf{C}(G)$ with the same "endpoints" and "direction." Next, let $\sim_\Sigma$ be the smallest congruence $\sim$ on $\mathbf{C}$ such that $g \sim g'$ for each equation $g = g'$ in Σ. Such a congruence exists simply because the intersection of a family of congruences is again a congruence. Taking the quotient by this congruence, we have a notion of a *finitely presented category*:

$$\mathbf{C}(G,\Sigma) = \mathbf{C}(G)/\!\sim_{\Sigma}$$

This is completely analogous to the notion of a finite presentation for groups, and indeed specializes to that notion in the case of a graph with only one vertex. The UMP of $\mathbf{C}(G,\Sigma)$ is then an obvious variant of that already given for groups.

Specifically, in $\mathbf{C}(G,\Sigma)$ there is a "diagram of type G," that is, a graph homomorphism $i : G \to |\mathbf{C}(G,\Sigma)|$, satisfying all the conditions $i(g) = i(g')$, for all $g = g' \in \Sigma$. Moreover, given any category $\mathbf{D}$ with a diagram of type G, say $h : G \to |\mathbf{D}|$, that satisfies all the conditions $h(g) = h(g')$, for all $g = g' \in \Sigma$, there is a unique functor $\bar{h} : \mathbf{C}(G,\Sigma) \to \mathbf{D}$ with $|\bar{h}| \circ i = h$.

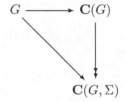

Just as in the case of presentations of groups, one can describe the construction of $\mathbf{C}(G,\Sigma)$ as a coequalizer for two functors. Indeed, suppose we have arrows $f, f' \in \mathbf{C}$. Take the least congruence $\sim$ on $\mathbf{C}$ with $f \sim f'$. Consider the diagram

$$\mathbf{C}(2) \underset{f'}{\overset{f}{\rightrightarrows}} \mathbf{C} \overset{q}{\longrightarrow} \mathbf{C}/\!\sim$$

where **2** is the graph with two vertices and an edge between them, f and f' are the unique functors taking the generating edge to the arrows by the same names, and q is the canonical functor to the quotient category. Then, q is a coequalizer of f and f'. To show this, take any $d : \mathbf{C} \to \mathbf{D}$ with

$$df = df'.$$

Since $\mathbf{C}(2)$ is free on the graph $\cdot \overset{x}{\to} \cdot$, and $f(x) = f$ and $f'(x) = f'$, we have

$$d(f) = d(f(x)) = d(f'(x)) = d(f').$$

Thus, $\langle f, f'\rangle \in \ker(d)$, so $\sim \subseteq \ker(d)$ (since $\sim$ is minimal with $f \sim f'$). So there is a functor $\bar{d} : \mathbf{C}/\!\sim\, \to \mathbf{D}$ such that $d = \bar{d} \circ q$ by the homomorphism theorem.

For the case of several equations rather than just one, in analogy with the case of finitely presented algebras (example 3.22), one replaces **2** by the graph $n \times \mathbf{2}$, and thus the free category $\mathbf{C}(2)$ by

$$\mathbf{C}(n \times \mathbf{2}) = n \times \mathbf{C}(\mathbf{2}) = \mathbf{C}(\mathbf{2}) + \cdots + \mathbf{C}(\mathbf{2}).$$

Example 4.15. The category with two uniquely isomorphic objects is not free on *any* graph, since it is finite, but has "loops" (cycles). But it *is* finitely presented with graph

$$A \underset{g}{\overset{f}{\rightleftarrows}} B$$

and relations

$$gf = 1_A, \qquad fg = 1_B.$$

Similarly, there are finitely presented categories with just one nonidentity arrow $f : \cdot \to \cdot$ and either

$$f \circ f = 1 \quad \text{or} \quad f \circ f = f.$$

In the first case, we have the group $\mathbb{Z}/2\mathbb{Z}$. In the second case, an "idempotent" (but not a group). Indeed, any of the cyclic groups

$$\mathbb{Z}_n \cong \mathbb{Z}/\mathbb{Z}n$$

occur in this way, with the graph $f : \star \to \star$ and the relation $f^n = 1$.

Of course, there are finitely presented categories with many objects as well. These are always given by a finite graph, the vertices of which are the objects and the edges of which generate the arrows, together with finitely many equations among paths of edges.

4.5   Exercises

1. Regarding a group G as a category with one object and every arrow an isomorphism, show that a categorical congruence $\sim$ on G is the same thing as (the equivalence relation on G determined by) a normal subgroup $N \subseteq G$, that is, show that the two kinds of things are in isomorphic correspondence.

   Show further that the quotient category $G/\sim$ and the factor group G/N coincide. Conclude that the homomorphism theorem for groups is a special case of the one for categories.

2. Consider the definition of a group in a category as applied to the category **Sets**$/I$ of sets sliced over a set I. Show that such a group G determines an I-indexed family of (ordinary) groups G_i by setting $G_i = G^{-1}(i)$ for each $i \in I$. Show that this determines a functor **Groups**(**Sets**$/I$) $\to$ **Groups**^{I} into the category of I-indexed families of groups and I-indexed families of homomorphisms.

3. Complete the proof that the groups in the category of groups are exactly the abelian groups by showing that any abelian group admits homomorphic group operations.

4. Use the Eckmann–Hilton argument to prove that every monoid in the category of groups is an internal group.

5. Given a homomorphism of abelian groups $f : A \to B$, define the cokernel $c : B \to C$ to be the quotient of B by the subgroup $\mathrm{im}(f) \subseteq B$.

   (a) Show that the cokernel has the following UMP: $c \circ f = 0$, and if $g : B \to G$ is any homomorphism with $g \circ f = 0$, then g factors uniquely through c as $g = u \circ c$.

   (b) Show that the cokernel is a particular kind of coequalizer, and use cokernels to construct arbitrary coequalizers.

   (c) Take the kernel of the cokernel, and show that $f : A \to B$ factors through it. Show, moreover, that this kernel is (isomorphic to) the image of $f : A \to B$. Infer that the factorization of $f : A \to B$ determined by cokernels agrees with that determined by taking the kernels.

6. Give four different presentations by generators and relations of the category **3**, pictured:

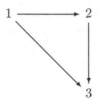

   Is **3** free?

7. Given a congruence $\sim$ on a category **C** and arrows in **C** as follows:

$$A \underset{f'}{\overset{f}{\rightrightarrows}} B \underset{g'}{\overset{g}{\rightrightarrows}} C$$

   show that $f \sim f'$ and $g \sim g'$ implies $g \circ f \sim g' \circ f'$.

8. Given functors $F, G : \mathbf{C} \to \mathbf{D}$ such that for all $C \in \mathbf{C}$, $FC = GC$, define a congruence on **D** by the condition

$$f \sim g \quad \text{iff} \quad \mathrm{dom}(f) = \mathrm{dom}(g) \ \& \ \mathrm{cod}(f) = \mathrm{cod}(g)$$

$$\& \ \forall \, \mathbf{E} \ \forall \, H : \mathbf{D} \to \mathbf{E} : \ HF = HG \Rightarrow H(f) = H(g)$$

   Prove that this is indeed a congruence. Prove, moreover, that $\mathbf{D}/\sim$ is the coequalizer of F and G.

9. Verify that the category $\mathbf{Ord}_{\mathrm{fin}}$ is indeed the free monoidal category on one object.

5

LIMITS AND COLIMITS

In this chapter, we first briefly discuss some topics—namely, subobjects and pullbacks—relating to the definitions that we already have. This is partly in order to see how these are used, but also because we need this material soon. Then we approach things more systematically, defining the general notion of a limit, which subsumes many of the particular abstract characterizations we have met so far. Of course, there is a dual notion of colimit, which also has many interesting applications. After a brief look at one more elementary notion in Chapter 6, we go on to what may be called "higher category theory."

5.1 Subobjects

We have seen that every subset $U \subseteq X$ of a set X occurs as an equalizer and that equalizers are always monomorphisms. Therefore, it is natural to regard monos as generalized subsets. That is, a mono in **Groups** can be regarded as a subgroup, a mono in **Top** as a subspace, and so on.

The rough idea is this: given a monomorphism,

$$m : M \rightarrowtail X$$

in a category **G** of structured sets of some sort—call them "gadgets"—the image sub*set*

$$\{m(y) \mid y \in M\} \subseteq X$$

which may be written as $m(M)$, is often a sub-gadget of X to which M is isomorphic via m.

$$m : M \xrightarrow{\sim} m(M) \subseteq X$$

More generally, we can think of the mono $m : M \rightarrowtail X$ itself as determining a "part" of X, even in categories that do not have underlying functions to take images of.

Definition 5.1. A *subobject* of an object X in a category **C** is a monomorphism:

$$m : M \rightarrowtail X.$$

Given subobjects m and m' of X, a morphism $f : m \to m'$ is an arrow in $\mathbf{C}/X$, as in

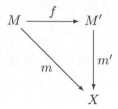

Thus, we have a category,

$$\mathrm{Sub}_{\mathbf{C}}(X)$$

of subobjects of X in $\mathbf{C}$.

In this definition, since m' is monic, there is at most one f as in the diagram above, so that $\mathrm{Sub}_{\mathbf{C}}(X)$ is a preorder category. We define the relation of *inclusion* of subobjects by

$$m \subseteq m' \quad \text{iff} \quad \text{there exists some } f : m \to m'$$

Finally, we say that m and m' are *equivalent*, written $m \equiv m'$, if and only if they are isomorphic as subobjects, that is, $m \subseteq m'$ and $m' \subseteq m$. This holds just if there are f and f' making both triangles below commute:

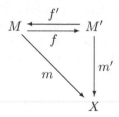

Observe that, in the above diagram, $m = m'f = mf'f$, and since m is monic, $f'f = 1_M$ and similarly $ff' = 1_{M'}$. So, $M \cong M'$ via f. Thus, we see that *equivalent subobjects have isomorphic domains.* We sometimes abuse notation and language by calling M the subobject when the mono $m : M \rightarrowtail X$ is clear.

Remark 5.2. It is often convenient to pass from the preorder

$$\mathrm{Sub}_{\mathbf{C}}(X)$$

to the *poset* given by factoring out the equivalence relation "$\equiv$". Then a subobject is an equivalence class of monos under mutual inclusion.

In **Sets**, under this notion of subobject, one then has an isomorphism,

$$\mathrm{Sub}_{\mathbf{Sets}}(X) \cong P(X)$$

that is, every subobject is represented by a unique subset. We shall use both notions of subobject, making clear when monos are intended, and when equivalence classes thereof are intended.

Note that if $M' \subseteq M$, then the arrow f which makes this so in

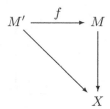

is also monic, so also M' is a subobject of M. Thus we have a functor

$$\mathrm{Sub}(M) \to \mathrm{Sub}(X)$$

defined by composition (since the composite of monos is monic).

In terms of generalized elements of an object X,

$$z : Z \to X$$

one can define a *local membership relation*,

$$z \in_X M$$

between such elements and subobjects $m : M \rightarrowtail X$ by

$$z \in_X M \quad \text{iff} \quad \text{there exists } f : Z \to M \text{ such that } z = mf.$$

Since m is monic, if z factors through it then it does so uniquely.

Example 5.3. An equalizer

$$E \longrightarrow A \underset{g}{\overset{f}{\rightrightarrows}} B$$

is a subobject of A with the property

$$z \in_A E \quad \text{iff} \quad f(z) = g(z).$$

Thus, we can regard E as the subobject of generalized elements $z : Z \to A$ such that $f(z) = g(z)$, suggestively,

$$E = \{ z \in Z \mid f(z) = g(z) \} \subseteq A.$$

In categorical logic, one develops a way of making this intuition even more precise by giving a calculus of such subobjects.

5.2 Pullbacks

The notion of a pullback, like that of a product, is one that comes up very often in mathematics and logic. It is a generalization of both intersection and inverse image.

We begin with the definition.

Definition 5.4. In any category $\mathbf{C}$, given arrows f, g with $\mathrm{cod}(f) = \mathrm{cod}(g)$,

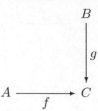

the *pullback* of f and g consists of arrows

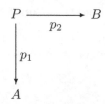

such that $fp_1 = gp_2$ and universal with this property. That is, given any $z_1 : Z \to A$ and $z_2 : Z \to B$ with $fz_1 = gz_2$, there exists a unique $u : Z \to P$ with $z_1 = p_1 u$ and $z_2 = p_2 u$. The situation is indicated in the following diagram:

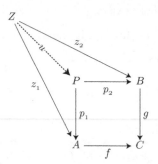

One sometimes uses product-style notation for pullbacks.

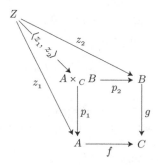

Pullbacks are clearly unique up to isomorphism since they are given by a universal mapping property (UMP). Here, this means that given two pullbacks of a given pair of arrows, the uniquely determined maps between the pullbacks are mutually inverse.

In terms of generalized elements, any $z \in A \times_C B$, can be written uniquely as $z = \langle z_1, z_2 \rangle$ with $f z_1 = g z_2$. This makes

$$A \times_C B = \{\langle z_1, z_2 \rangle \in A \times B \mid f z_1 = g z_2\}$$

look like a subobject of $A \times B$, determined as an equalizer of $f \circ \pi_1$ and $g \circ \pi_2$. In fact, this is so.

Proposition 5.5. *In a category with products and equalizers, given a corner of arrows*

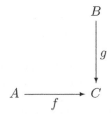

Consider the diagram

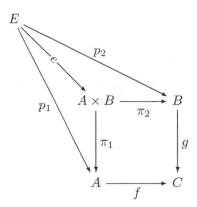

in which e is an equalizer of $f\pi_1$ and $g\pi_2$ and $p_1 = \pi_1 e$, $p_2 = \pi_2 e$. Then, E, p_1, p_2 is a pullback of f and g. Conversely, if E, p_1, p_2 are given as such a pullback, then the arrow

$$e = \langle p_1, p_2 \rangle : E \to A \times B$$

is an equalizer of $f\pi_1$ and $g\pi_2$.

Proof. Take

$$Z \xrightarrow{\ z_2\ } B$$
$$\downarrow{\scriptstyle z_1}$$
$$A$$

with $fz_1 = gz_2$. We have $\langle z_1, z_2 \rangle : Z \to A \times B$, so

$$f\pi_1 \langle z_1, z_2 \rangle = g\pi_2 \langle z_1, z_2 \rangle.$$

Thus, there is a $u : Z \to E$ to the equalizer with $eu = \langle z_1, z_2 \rangle$. Then,

$$p_1 u = \pi_1 eu = \pi_1 \langle z_1, z_2 \rangle = z_1$$

and

$$p_2 u = \pi_2 eu = \pi_2 \langle z_1, z_2 \rangle = z_2.$$

If also $u' : Z \to E$ has $p_i u' = z_i, i = 1, 2$, then $\pi_i eu' = z_i$ so $eu' = \langle z_1, z_2 \rangle = eu$ whence $u' = u$ since e in monic. The converse is similar. $\qquad\square$

Corollary 5.6. *If a category* **C** *has binary products and equalizers, then it has pullbacks.*

The foregoing gives an explicit construction of a pullback in **Sets** as a subset of the product:

$$\{\langle a, b \rangle \mid fa = gb\} = A \times_C B \hookrightarrow A \times B$$

Example 5.7. In **Sets**, take a function $f : A \to B$ and a subset $V \subseteq B$. Let, as usual,

$$f^{-1}(V) = \{a \in A \mid f(a) \in V\} \subseteq A$$

and consider

$$
\begin{array}{ccc}
f^{-1}(V) & \xrightarrow{\ \bar{f}\ } & V \\
{\scriptstyle j}\downarrow & & \downarrow{\scriptstyle i} \\
A & \xrightarrow[\ f\ ]{} & B
\end{array}
$$

where i and j are the canonical inclusions and $\bar{f}$ is the evident factorization of the restriction of f to $f^{-1}(V)$ (since $a \in f^{-1}(V) \Rightarrow f(a) \in V$).

This diagram is a pullback (observe that $z \in f^{-1}(V) \Leftrightarrow fz \in V$ for all $z : Z \to A$). Thus, the inverse image

$$f^{-1}(V) \subseteq A$$

is *determined uniquely up to isomorphism* as a pullback.

As suggested by the previous example, we can use pullbacks to *define* inverse images in categories other than **Sets**. Indeed, given a pullback in any category

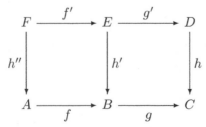

if m is monic, then m' is monic. (Exercise!)

Thus, we see that, for fixed $f : A \to B$, taking pullbacks induces a map

$$f^{-1} : \mathrm{Sub}(B) \to \mathrm{Sub}(A)$$

$$m \mapsto m'$$

We show that f^{-1} also respects equivalence of subobjects,

$$M \equiv N \Rightarrow f^{-1}(M) \equiv f^{-1}(N)$$

by showing that f^{-1} is a functor, which is our next goal.

5.3 Properties of pullbacks

We start with the following simple lemma, which seems to come up all the time.

Lemma 5.8. *(Two-pullbacks) Consider the commutative diagram below in a category with pullbacks:*

$$
\begin{array}{ccccc}
F & \xrightarrow{f'} & E & \xrightarrow{g'} & D \\
{\scriptstyle h''}\downarrow & & {\scriptstyle h'}\downarrow & & \downarrow{\scriptstyle h} \\
A & \xrightarrow{f} & B & \xrightarrow{g} & C
\end{array}
$$

1. *If the two squares are pullbacks, so is the outer rectangle. Thus,*

$$A \times_B (B \times_C D) \cong A \times_C D.$$

2. *If the right square and the outer rectangle are pullbacks, so is the left square.*

Proof. Diagram chase.   □

Corollary 5.9. *The pullback of a commutative triangle is a commutative trian-gle. Specifically, given a commutative triangle as on the right end of the following "prism diagram":*

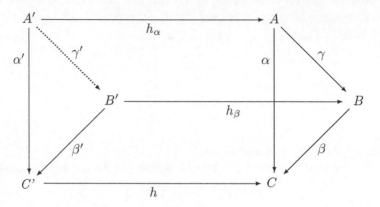

for any $h : C' \to C$, if one can form the pullbacks α' and β' as on the left end, then there exists a unique γ' as indicated, making the left end a commutative triangle, and the upper face a commutative rectangle, and indeed a pullback.

Proof. Apply the two-pullbacks lemma.   □

Proposition 5.10. *Pullback is a functor. That is, for fixed $h : C' \to C$ in a category $\mathbf{C}$ with pullbacks, there is a functor*

$$h^* : \mathbf{C}/C \to \mathbf{C}/C'$$

defined by

$$(A \xrightarrow{\alpha} C) \mapsto (C' \times_C A \xrightarrow{\alpha'} C')$$

where α' is the pullback of α along h, and the effect on an arrow $\gamma : \alpha \to \beta$ is given by the foregoing corollary.

Proof. One must check that

$$h^*(1_X) = 1_{h^*X}$$

and

$$h^*(g \circ f) = h^*(g) \circ h^*(f).$$

These can easily be verified by repeated applications of the two-pullbacks lemma. For example, for the first condition, consider

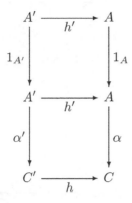

If the lower square is a pullback, then plainly so is the outer rectangle, whence the upper square is, too, and we have

$$h^* 1_\alpha = 1_{\alpha'} = 1_{h^*\alpha}.$$

□

Corollary 5.11. *Let* **C** *be a category with pullbacks. For any arrow* $f : A \to B$ *in* **C***, we have the following diagram of categories and functors:*

$$
\begin{array}{ccc}
\mathrm{Sub}(A) & \xleftarrow{\ f^{-1}\ } & \mathrm{Sub}(B) \\
\downarrow & & \downarrow \\
\mathbf{C}/A & \xleftarrow{\ f^*\ } & \mathbf{C}/B
\end{array}
$$

This commutes simply because f^{-1} *is defined to be the restriction of* f^* *to the subcategory* $\mathrm{Sub}(B)$*. Thus, in particular,* f^{-1} *is functorial:*

$$M \subseteq N \ \Rightarrow\ f^{-1}(M) \subseteq f^{-1}(N)$$

It follows that $M \equiv N$ *implies* $f^{-1}(M) \equiv f^{-1}(N)$*, so that* f^{-1} *is also defined on equivalence classes.*

$$f^{-1}/\equiv\ :\ \mathrm{Sub}(B)/\equiv\ \longrightarrow\ \mathrm{Sub}(A)/\equiv$$

Example 5.12. Consider a pullback in **Sets**:

$$
\begin{array}{ccc}
E & \xrightarrow{\;f'\;} & B \\
{\scriptstyle g'}\big\downarrow & & \big\downarrow{\scriptstyle g} \\
A & \xrightarrow[f]{} & C
\end{array}
$$

We saw that

$$E = \{\langle a, b\rangle \mid f(a) = g(b)\}$$

can be constructed as an equalizer

$$
E \xrightarrow{\;\langle f', g'\rangle\;} A \times B
\begin{array}{c} \xrightarrow{f\pi_1} \\[-4pt] \xrightarrow[g\pi_2]{} \end{array}
C
$$

Now let $B = 1$, $C = 2 = \{\top, \bot\}$, and $g = \top : 1 \to 2$. Then, the equalizer

$$
E \xrightarrow{\quad\quad} A \times 1
\begin{array}{c} \xrightarrow{f\pi_1} \\[-4pt] \xrightarrow[\top\pi_2]{} \end{array}
2
$$

is how we already described the "extension" of the "propositional function" $f : A \to 2$. Therefore, we can rephrase the correspondence between subsets $U \subseteq A$ and their characteristic functions $\chi_U : A \to 2$ in terms of pullbacks:

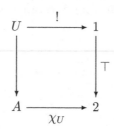

Precisely, the isomorphism,

$$2^A \cong P(A)$$

given by taking a function $\varphi : A \to 2$ to its "extension"

$$V_\varphi = \{x \in A \mid \varphi(x) = \top\}$$

can be described as a pullback.

$$V_\varphi = \{x \in A \mid \varphi(x) = \top\} = \varphi^{-1}(\top)$$

Now suppose we have any function

$$f : B \to A$$

and consider the induced inverse image operation

$$f^{-1} : P(A) \to P(B)$$

given by pullback, as in example 5.9 above. Taking the extension $V_\varphi \subseteq A$, consider the two-pullbacks diagram:

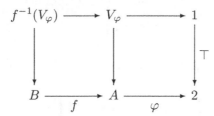

We therefore have (by the two-pullbacks lemma)

$$f^{-1}(V_\varphi) = f^{-1}(\varphi^{-1}(\top)) = (\varphi f)^{-1}(\top) = V_{\varphi f}$$

which from a logical point of view expresses the fact that the substitution of a term f for the variable x over A in the propositional function φ is modeled by taking the pullback along f of the corresponding extension

$$f^{-1}(\{x \in A \mid \varphi(x) = \top\}) = \{y \in B \mid \varphi(f(y)) = \top\}.$$

Note that we have shown that for any function $f : B \to A$ the following square commutes:

$$
\begin{array}{ccc}
2^A & \xrightarrow{\cong} & P(A) \\
{\scriptstyle 2^f}\big\downarrow & & \big\downarrow{\scriptstyle f^{-1}} \\
2^B & \xrightarrow{\cong} & P(B)
\end{array}
$$

where $2^f : 2^A \to 2^B$ is precomposition $2^f(g) = g \circ f$. In a situation like this, one says that the isomorphism

$$2^A \cong P(A)$$

is *natural* in A, which is obviously a much stronger condition than just having isomorphisms at each object A. We will consider such "naturality" systematically later. It was in fact one of the phenomena that originally gave rise to category theory.

Example 5.13. Let I be an index set, and consider an I-indexed family of sets:

$$(A_i)_{i \in I}$$

Given any function $\alpha : J \to I$, there is a J-indexed family

$$(A_{\alpha(j)})_{j \in J} ,$$

obtained by "reindexing along α." This reindexing can also be described as a pullback. Specifically, for each set A_i take the constant, i-valued function $p_i : A_i \to I$ and consider the induced map on the coproduct

$$p = [p_i] : \coprod_{i \in I} A_i \to I$$

The reindexed family $(A_{\alpha(j)})_{j \in J}$ can be obtained by taking a pullback along α, as indicated in the following diagram:

$$
\begin{array}{ccc}
\coprod_{j \in J} A_{\alpha(j)} & \longrightarrow & \coprod_{i \in I} A_i \\
\downarrow q & & \downarrow p \\
J & \xrightarrow{\;\;\alpha\;\;} & I
\end{array}
$$

where q is the indexing projection for $(A_{\alpha(j)})_{j \in J}$ analogous to p. In other words, we have

$$J \times_I \left(\coprod_{i \in I} A_i \right) \cong \coprod_{j \in J} A_{\alpha(j)}$$

The reader should work out the details as an instructive exercise.

5.4 Limits

We have already seen that the notions of product, equalizer, and pullback are not independent; the precise relation between them is this.

Proposition 5.14. *A category has finite products and equalizers iff it has pullbacks and a terminal object.*

Proof. The "only if" direction has already been done. For the other direction, suppose **C** has pullbacks and a terminal object 1.

- For any objects A, B we clearly have $A \times B \cong A \times_1 B$, as indicated in the following:

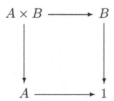

- For any arrows $f, g : A \to B$, the equalizer $e : E \to A$ is constructed as the following pullback:

$$
\begin{array}{ccc}
E & \xrightarrow{\ h\ } & B \\
\downarrow{\scriptstyle e} & & \downarrow{\scriptstyle \Delta = \langle 1_B, 1_B \rangle} \\
A & \xrightarrow[\langle f,g \rangle]{} & B \times B
\end{array}
$$

In terms of generalized elements,

$$E = \{(a,b) \mid \langle f, g \rangle(a) = \Delta b\}$$

where $\langle f, g \rangle(a) = \langle fa, ga \rangle$ and $\Delta(b) = \langle b, b \rangle$. So,

$$E = \{\langle a, b \rangle \mid f(a) = b = g(a)\}$$
$$\cong \{a \mid f(a) = g(a)\}$$

which is just what we want. An easy diagram chase shows that

$$E \xrightarrow{\ e\ } A \underset{g}{\overset{f}{\rightrightarrows}} B$$

is indeed an equalizer.   □

Product, terminal object, pullback, and equalizer, are all special cases of the general notion of a *limit*, which we consider now. First, we need some preliminary definitions.

Definition 5.15. Let **J** and **C** be categories. A *diagram* of *type* **J** in **C** is a functor.

$$D : \mathbf{J} \to \mathbf{C}.$$

We write the objects in the "index category" **J** lower case, $i, j, \ldots$ and the values of the functor $D : \mathbf{J} \to \mathbf{C}$ in the form D_i, D_j, etc.

A *cone* to a diagram D consists of an object C in **C** and a family of arrows in **C**,

$$c_j : C \to D_j$$

one for each object $j \in J$, such that for each arrow $\alpha : i \to j$ in **J**, the following triangle commutes:

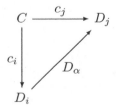

A *morphism* of cones

$$\vartheta : (C, c_j) \to (C', c'_j)$$

is an arrow ϑ in **C** making each triangle,

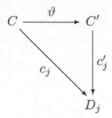

commute. That is, such that $c_j = c'_j \circ \vartheta$ for all $j \in \mathbf{J}$. Thus, we have an evident category

$$\mathbf{Cone}(D)$$

of cones to D.

We are here thinking of the diagram D as a "picture of **J** in **C**." A cone to such a diagram D is then imagined as a many-sided pyramid over the "base" D and a morphism of cones is an arrow between the apexes of such pyramids. (The reader should draw some pictures at this point!)

Definition 5.16. A *limit* for a diagram $D : \mathbf{J} \to \mathbf{C}$ is a terminal object in **Cone**(D). A *finite limit* is a limit for a diagram on a finite index category **J**.

We often denote a limit in the form

$$p_i : \varprojlim_j D_j \to D_i.$$

Spelling out the definition, the limit of a diagram D has the following UMP: given any cone (C, c_j) to D, there is a unique arrow $u : C \to \varprojlim_j D_j$ such that for all j,

$$p_j \circ u = c_j.$$

Thus, the limiting cone $(\varprojlim_j D_j, p_j)$ can be thought of as the "closest" cone to the diagram D, and indeed any other cone (C, c_j) comes from it just by composing with an arrow at the vertex, namely $u : C \to \varprojlim_j D_j$.

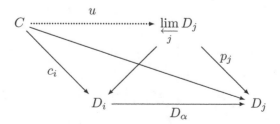

Example 5.17. Take $\mathbf{J} = \{1, 2\}$ the discrete category with two objects and no nonidentity arrows. A diagram $D : \mathbf{J} \to \mathbf{C}$ is a pair of objects $D_1, D_2 \in \mathbf{C}$. A cone on D is an object of $\mathbf{C}$ equipped with arrows

$$D_1 \xleftarrow{\quad c_1 \quad} C \xrightarrow{\quad c_2 \quad} D_2.$$

And a limit of D is a terminal such cone, that is, a *product* in $\mathbf{C}$ of D_1 and D_2,

$$D_1 \xleftarrow{\quad p_1 \quad} D_1 \times D_2 \xrightarrow{\quad p_2 \quad} D_2.$$

Thus, in this case,

$$\varprojlim_j D_j \cong D_1 \times D_2.$$

Example 5.18. Take $\mathbf{J}$ to be the following category:

$$\cdot \underset{\beta}{\overset{\alpha}{\rightrightarrows}} \cdot$$

A diagram of type $\mathbf{J}$ looks like

$$D_1 \underset{D_\beta}{\overset{D_\alpha}{\rightrightarrows}} D_2$$

and a cone is a pair of arrows

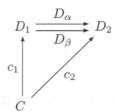

such that $D_\alpha c_1 = c_2$ and $D_\beta c_1 = c_2$; thus, $D_\alpha c_1 = D_\beta c_1$. A limit for D is therefore an *equalizer* for D_α, D_β.

Example 5.19. If $\mathbf{J}$ is empty, there is just one diagram $D : \mathbf{J} \to \mathbf{C}$, and a limit for it is thus a *terminal object* in $\mathbf{C}$,

$$\varprojlim_{j \in \mathbf{0}} D_j \cong 1.$$

Example 5.20. If $\mathbf{J}$ is the finite category

we see that a limit for a diagram of the form

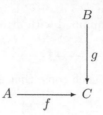

is just a pullback of f and g,

$$\varprojlim_{j} D_j \;\cong\; A \times_C B.$$

Thus, we have shown half of the following.

Proposition 5.21. *A category has all finite limits iff it has finite products and equalizers (resp. pullbacks and a terminal object by the last proposition).*

Here, a category **C** is said to *have all finite limits* if every finite diagram $D :$ **J** $\to$ **C** has a limit in **C**.

Proof. We need to show that any finite limit can be constructed from finite products and equalizers. Take a finite diagram

$$D : \mathbf{J} \to \mathbf{C}.$$

As a first approximation, the product

$$\prod_{i \in \mathbf{J}_0} D_i \tag{5.1}$$

over the set $\mathbf{J}_0$ of objects at least has projections $p_j : \coprod_{i \in \mathbf{J}_0} D_i \to D_j$ of the right sort. But these cannot be expected to commute with the arrows $D_\alpha : D_i \to D_j$ in the diagram D, as they must. So, as in making a pullback from a product and an equalizer, we consider also the product $\prod_{(\alpha : i \to j) \in \mathbf{J}_1} D_j$ over all the arrows (the set $\mathbf{J}_1$), and two special maps,

$$\prod_{i} D_i \;\overset{\phi}{\underset{\psi}{\rightrightarrows}}\; \prod_{\alpha : i \to j} D_j$$

which record the effect of the arrows in the diagram on the product of the objects. Specifically, we define ϕ and ψ by taking their composites with the projections π_α from the second product to be, respectively,

$$\pi_\alpha \circ \phi = \phi_\alpha = \pi_{\mathrm{cod}(\alpha)}$$
$$\pi_\alpha \circ \psi = \psi_\alpha = D_\alpha \circ \pi_{\mathrm{dom}(\alpha)}$$

where $\pi_{\mathrm{cod}(\alpha)}$ and $\pi_{\mathrm{dom}(\alpha)}$ are projections from the first product.

Now, in order to get the subobject of the product 5.1 on which the arrows in the diagram D commute, we take the equalizer:

$$E \xrightarrow{\ e\ } \prod_i D_i \ \underset{\psi}{\overset{\phi}{\rightrightarrows}} \ \prod_{\alpha:i\to j} D_j$$

We show that (E, e_i) is a limit for D, where $e_i = \pi_i \circ e$. To that end, take any arrow $c : C \to \prod_i D_i$, and write $c = \langle c_i \rangle$ for $c_i = \pi_i \circ c$. Observe that the family of arrows $(c_i : C \to D_i)$ is a cone to D if and only if $\phi c = \psi c$. Indeed,

$$\phi\langle c_i \rangle = \psi\langle c_i \rangle$$

iff for all α,

$$\pi_\alpha \phi\langle c_i \rangle = \pi_\alpha \psi\langle c_i \rangle.$$

But,

$$\pi_\alpha \phi\langle c_i \rangle = \phi_\alpha\langle c_i \rangle = \pi_{\mathrm{cod}(\alpha)}\langle c_i \rangle = c_j$$

and

$$\pi_\alpha \psi\langle c_i \rangle = \psi_\alpha\langle c_i \rangle = D_\alpha \circ \pi_{\mathrm{dom}(\alpha)}\langle c_i \rangle = D_\alpha \circ c_i.$$

Whence $\phi c = \psi c$ iff for all $\alpha : i \to j$ we have $c_j = D_\alpha \circ c_i$ thus, iff $(c_i : C \to D_i)$ is a cone, as claimed. It follows that (E, e_i) is a cone, and that any cone $(c_i : C \to D_i)$ gives an arrow $\langle c_i \rangle : C \to \prod_i D_i$ with $\phi\langle c_i \rangle = \psi\langle c_i \rangle$, thus there is a unique factorization $u : C \to E$ of $\langle c_i \rangle$ through E, which is clearly a morphism of cones. $\qquad\square$

Since we made no real use of the finiteness of the index category apart from the existence of certain products, essentially the same proof yields the following.

Corollary 5.22. *A category has all limits of some cardinality iff it has all equalizers and products of that cardinality, where* **C** *is said to have limits (resp. products) of cardinality κ iff* **C** *has a limit for every diagram $D : \mathbf{J} \to \mathbf{C}$, where* $\mathrm{card}(\mathbf{J}_1) \le \kappa$ *(resp.* **C** *has all products of κ many objects).*

The notions of cones and limits of course dualize to give those of *cocones* and *colimits*. One then has the following dual theorem.

Theorem 5.23. *A category* **C** *has finite colimits iff it has finite coproducts and coequalizers (resp. iff it has pushouts and an initial object).* **C** *has all colimits of size κ iff it has coequalizers and coproducts of size κ.*

5.5 Preservation of limits

Here is an application of the construction of limits by products and equalizers.

Definition 5.24. A functor $F : \mathbf{C} \to \mathbf{D}$ is said to *preserve limits of type* $\mathbf{J}$ if, whenever $p_j : L \to D_j$ is a limit for a diagram $D : \mathbf{J} \to \mathbf{C}$; the cone $Fp_j : FL \to FD_j$ is then a limit for the diagram $FD : \mathbf{J} \to \mathbf{D}$. Briefly,

$$F(\varprojlim D_j) \cong \varprojlim F(D_j).$$

A functor that preserves all limits is said to be *continuous*.

For example, let $\mathbf{C}$ be a locally small category with all small limits, such as posets or monoids. Recall the representable functor

$$\mathrm{Hom}(C, -) : \mathbf{C} \to \mathbf{Sets}$$

for any object $C \in \mathbf{C}$, taking $f : X \to Y$ to

$$f_* : \mathrm{Hom}(C, X) \to \mathrm{Hom}(C, Y)$$

where $f_*(g : C \to X) = f \circ g$.

Proposition 5.25. *The representable functors* $\mathrm{Hom}(C, -)$ *preserve all limits.*

Since limits in $\mathbf{C}$ can be constructed from products and equalizers, it suffices to show that $\mathrm{Hom}(C, -)$ preserves products and equalizers. (Actually, even if $\mathbf{C}$ does not have all limits, the representable functors will preserve those limits that do exist; we leave that as an exercise.)

Proof.   • $\mathbf{C}$ has a terminal object 1, for which,

$$\mathrm{Hom}(C, 1) = \{!_C\} \cong 1.$$

• Consider a binary product $X \times Y$ in $\mathbf{C}$. Then, we already know that

$$\mathrm{Hom}(C, X \times Y) \cong \mathrm{Hom}(C, X) \times \mathrm{Hom}(C, Y)$$

by composing any $f : C \to X \times Y$ with the two product projections $p_1 : X \times Y \to X$ and $p_2 : X \times Y \to Y$.

• For arbitrary products $\prod_{i \in I} X_i$, one has analogously

$$\mathrm{Hom}(C, \prod_i X_i) \cong \prod_i \mathrm{Hom}(C, X_i)$$

• Given an equalizer in $\mathbf{C}$,

$$E \xrightarrow{\quad e \quad} X \overset{f}{\underset{g}{\rightrightarrows}} Y$$

consider the resulting diagram:

$$\mathrm{Hom}(C, E) \xrightarrow{\quad e_* \quad} \mathrm{Hom}(C, X) \overset{f_*}{\underset{g_*}{\rightrightarrows}} \mathrm{Hom}(C, Y).$$

To show this is an equalizer in **Sets**, let $h : C \rightarrow X \in \mathrm{Hom}(C, X)$ with $f_* h = g_* h$. Then $fh = gh$, so there is a unique $u : C \rightarrow E$ such that $eu = h$. Thus, we have a unique $u \in \mathrm{Hom}(C, E)$ with $e_* u = eu = h$. So, $e_* : \mathrm{Hom}(C, E) \rightarrow \mathrm{Hom}(C, X)$ is indeed the equalizer of f_* and g_*.

$\square$

Definition 5.26. A functor of the form $F : \mathbf{C}^{\mathrm{op}} \rightarrow \mathbf{D}$ is called a *contravariant functor* on $\mathbf{C}$. Explicitly, such a functor takes $f : A \rightarrow B$ to $F(f) : F(B) \rightarrow F(A)$ and $F(g \circ f) = F(f) \circ F(g)$.

A typical example of a contravariant functor is a representable functor of the form,

$$\mathrm{Hom}_{\mathbf{C}}(-, C) : \mathbf{C}^{\mathrm{op}} \rightarrow \mathbf{Sets}$$

for any $C \in \mathbf{C}$ (where $\mathbf{C}$ is any locally small category). Such a contravariant representable functor takes $f : X \rightarrow Y$ to

$$f^* : \mathrm{Hom}(Y, C) \rightarrow \mathrm{Hom}(X, C)$$

by $f^*(g : X \rightarrow C) = g \circ f$.

Then, the following is the dual version of the foregoing proposition.

Corollary 5.27. *Contravariant representable functors map all colimits to limits.*

For example, given a coproduct $X + Y$ in any locally small category $\mathbf{C}$, there is a canonical isomorphism,

$$\mathrm{Hom}(X + Y, C) \cong \mathrm{Hom}(X, C) \times \mathrm{Hom}(Y, C) \tag{5.2}$$

given by precomposing with the two coproduct inclusions.

From an example in Section 2.3, we can therefore conclude that the ultrafilters in a coproduct $A + B$ of Boolean algebras correspond exactly to pairs of ultrafilters (U, V), with U in A and V in B. This follows because we showed there that the ultrafilter functor $\mathrm{Ult} : \mathbf{BA}^{\mathrm{op}} \rightarrow \mathbf{Sets}$ is representable:

$$\mathrm{Ult}(B) \cong \mathrm{Hom}_{\mathbf{BA}}(B, 2).$$

Another case of the above iso (5.2) is the familiar law of exponents for sets:

$$C^{X+Y} \cong C^X \times C^Y$$

The arithmetical law of exponents $k^{m+n} = k^n \cdot k^m$ is actually a special case of this.

5.6   Colimits

Let us briefly discuss some special colimits, since we did not really say much about them Section 5.5.

First, we consider *pushouts* in **Sets**. Suppose we have two functions

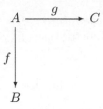

We can construct the pushout of f and g like this. Start with the coproduct (disjoint sum):

$$B \longrightarrow B + C \longleftarrow C$$

Now identify those elements $b \in B$ and $c \in C$ such that, for some $a \in A$,

$$f(a) = b \quad \text{and} \quad g(a) = c$$

That is, we take the equivalence relation $\sim$ on $B + C$ generated by the conditions $f(a) \sim g(a)$ for all $a \in A$.

Finally, we take the quotient by $\sim$ to get the pushout

$$(B + C)/\sim \quad \cong \quad B +_A C,$$

which can be imagined as B placed next to C, with the respective parts that are images of A "pasted together" or overlapping. This construction follows simply by dualizing the one for pullbacks by products and equalizers.

Example 5.28. Pushouts in **Top** are similarly formed from coproducts and coequalizers, which can be made first in **Sets** and then topologized as sum and quotient spaces. Pushouts are used, for example, to construct spheres from disks. Indeed, let D^2 be the (two-dimensional) disk and S^1 the one-dimensional sphere (i.e., the circle), with its inclusion $i : S^1 \to D^2$ as the boundary of the disk. Then, the two-sphere S^2 is the pushout,

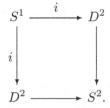

Can you see the analogous construction of S^1 at the next lower dimension?

In general, a colimit for a diagram $D : \mathbf{J} \to \mathbf{C}$ is, of course, an initial object in the category of *cocones*. Explicitly, a *cocone from the base D* consists of an object C (the vertex) and arrows $c_j : D_j \to C$ for each $j \in \mathbf{J}$, such that for all $\alpha : i \to j$ in $\mathbf{J}$,

$$c_j \circ D(\alpha) = c_i$$

A morphism of cocones $f : (C, (c_j)) \to (C', (c_j'))$ is an arrow $f : C \to C'$ in $\mathbf{C}$ such that $f \circ c_j = c_j'$ for all $j \in \mathbf{J}$. An initial cocone is the expected thing: one that maps uniquely *to* any other cocone from D. We write such a colimit in the form

$$\varinjlim_{j \in \mathbf{J}} D_j$$

Now let us consider some examples of a particular kind of colimit that comes up quite often, namely over a linearly ordered index category. Our first example is what is sometimes called a *direct limit* of a sequence of algebraic objects, say groups. A similar construction works for any sort of algebras (but non-equational conditions are not always preserved by direct limits).

Example 5.29. *Direct limit of groups.* Suppose we are given a sequence,

$$G_0 \xrightarrow[g_0]{} G_1 \xrightarrow[g_1]{} G_2 \xrightarrow[g_2]{} \cdots$$

of groups and homomorphisms, and we want a "colimiting" group G_∞ with homomorphisms

$$u_n : G_n \to G_\infty$$

satisfying $u_{n+1} \circ g_n = u_n$. Moreover, G_∞ should be "universal" with this property. I think you can see the colimit setup here:

- the index category is the ordinal number $\omega = (\mathbb{N}, \leq)$, regarded as a poset category,
- the sequence

$$G_0 \xrightarrow[g_0]{} G_1 \xrightarrow[g_1]{} G_2 \xrightarrow[g_2]{} \cdots$$

  is a diagram of type ω in the category **Groups**,
- the colimiting group is the colimit of the sequence

$$G_\infty \cong \varinjlim_{n \in \omega} G_n$$

This group always exists, and can be constructed as follows. Begin with the coproduct (disjoint sum) of sets

$$\coprod_{n \in \omega} G_n.$$

Then make identifications $x_n \sim y_m$, where $x_n \in G_n$ and $y_m \in G_m$, to ensure in particular that

$$x_n \sim g_n(x_n)$$

for all $x_n \in G_n$ and $g_n : G_n \to G_{n+1}$.

This means, specifically, that the elements of G_∞ are equivalence classes of the form

$$[x_n], \quad x_n \in G_n$$

for any n, and $[x_n] = [y_m]$ iff for some $k \geq m, n$,

$$g_{n,k}(x_n) = g_{m,k}(y_m)$$

where, generally, if $i \leq j$, we define

$$g_{i,j} : G_i \to \cdots \to G_j$$

by composing consecutive g's as in $g_{i,j} = g_{j-1} \circ \ldots \circ g_i$. The reader can easily check that this is indeed the equivalence relation generated by all the conditions $x_n \sim g_n(x_n)$.

The operations on G_∞ are now defined by

$$[x] \cdot [y] = [x' \cdot y']$$

where $x \sim x'$, $y \sim y'$, and $x', y' \in G_n$ for n sufficiently large. The unit is just $[u_0]$, and we take,

$$[x]^{-1} = [x^{-1}].$$

One can easily check that these operations are well defined, and determine a group structure on G_∞, which moreover makes all the evident functions

$$u_n : G_n \to G_\infty , \qquad u_n(x) = [x]$$

into homomorphisms.

The universality of G_∞ and the u_n results from the fact that the construction is essentially a colimit in **Sets**, equipped with an induced group structure. Indeed, given any group H and homomorphisms $h_n : G_n \to H$ with $h_{n+1} \circ g_n = h_n$ define $h_\infty : G_\infty \to H$ by $h_\infty([x_n]) = h_n(x_n)$. This is easily seen to be well defined and indeed a homomorphism. Moreover, it is the unique *function* that commutes with all the u_n.

The fact that the ω-colimit G_∞ of groups can be constructed as the colimit of the underlying sets is a case of a general phenomenon, expressed by saying that the forgetful functor $U : \textbf{Groups} \to \textbf{Sets}$ "creates ω-colimits."

Definition 5.30. A functor $F : \textbf{C} \to \textbf{D}$ is said to *create limits of type* **J** if for every diagram $C : \textbf{J} \to \textbf{C}$ and limit $p_j : L \to FC_j$ in **D** there is a unique cone $\overline{p_j} : \overline{L} \to C_j$ in **C** with $F(\overline{L}) = L$ and $F(\overline{p_j}) = p_j$, which, furthermore, is a limit for C. Briefly, every limit in $\mathbf{D}$ is the image of a unique cone in $\mathbf{C}$, which is a limit there. The notion of *creating colimits* is defined analogously.

In these terms, then, we have the following proposition, the remaining details of which have in effect already been shown.

Proposition 5.31. *The forgetful functor* U : **Groups** $\rightarrow$ **Sets** *creates ω-colimits. It also creates all limits.*

The same fact holds quite generally for other categories of algebraic objects, that is, sets equipped with operations satisfying some equations. Observe that not *all* colimits are created in this way. For instance, we have already seen (in proposition 3.11) that the coproduct of two abelian groups has their *product* as underlying set.

Example 5.32. *Cumulative hierarchy.* Another example of an ω-colimit is the "cumulative hierarchy" construction encountered in set theory. Let us set

$$V_0 = \emptyset$$

$$V_1 = \mathcal{P}(\emptyset)$$

$$\vdots$$

$$V_{n+1} = \mathcal{P}(V_n)$$

Then there is a sequence of subset inclusions,

$$\emptyset = V_0 \subseteq V_1 \subseteq V_2 \subseteq \cdots$$

since, generally, $A \subseteq B$ implies $\mathcal{P}(A) \subseteq \mathcal{P}(B)$ for any sets A and B. The colimit of the sequence

$$V_\omega = \varinjlim_n V_n$$

is called the *cumulative hierarchy* of rank ω. One can, of course, continue this construction through higher ordinals $\omega + 1, \omega + 2, \ldots$.

More generally, let us start with some set A (of "atoms"), and let

$$V_0(A) = A$$

and then put

$$V_{n+1}(A) = A + \mathcal{P}(V_n(A)),$$

that is, $V_1(A) = A + \mathcal{P}(A)$ is the set of all elements *and* subsets of A. There is a sequence $V_0(A) \rightarrow V_1(A) \rightarrow V_2(A) \rightarrow \ldots$ as follows. Let

$$v_0 : V_0(A) = A \rightarrow A + \mathcal{P}(A) = V_1(A)$$

be the left coproduct inclusion. Given $v_{n-1} : V_{n-1}(A) \rightarrow V_n(A)$, let $v_n : V_n(A) \rightarrow V_{n+1}(A)$ be defined by

$$v_n = 1_A + \mathcal{P}_!(v_{n-1}) : A + \mathcal{P}(V_{n-1}(A)) \rightarrow A + \mathcal{P}(V_n(A))$$

where $\mathcal{P}_!$ denotes the *covariant* powerset functor, taking a function $f : X \to Y$ to the "image under f" operation $\mathcal{P}_!(f) : \mathcal{P}(X) \to \mathcal{P}(Y)$, defined by taking $U \subseteq X$ to

$$\mathcal{P}_!(f)(U) = \{f(u) \mid u \in U\} \subseteq Y.$$

The idea behind the sequence is that we start with A, add all the subsets of A, then add all the *new* subsets that can be formed from all of those elements, and so on. The colimit of the sequence

$$V_\omega(A) = \varinjlim_n V_n(A)$$

is called the *cumulative hierarchy* (of rank ω) over A. Of course, $V_\omega = V_\omega(\emptyset)$.

Now suppose we have some function

$$f : A \to B.$$

Then, there is a map

$$V_\omega(f) : V_\omega(A) \to V_\omega(B),$$

determined by the colimit description of V_ω, as indicated in the following diagram:

$$
\begin{array}{ccccccccc}
V_0(A) & \longrightarrow & V_1(A) & \longrightarrow & V_2(A) & \longrightarrow & \cdots & \longrightarrow & V_\omega(A) \\
\downarrow{\scriptstyle f_0} & & \downarrow{\scriptstyle f_1} & & \downarrow{\scriptstyle f_2} & & \cdots & & \downarrow{\scriptstyle f_\omega} \\
V_0(B) & \longrightarrow & V_1(B) & \longrightarrow & V_2(B) & \longrightarrow & \cdots & \longrightarrow & V_\omega(B)
\end{array}
$$

Here, the f_n are defined by

$$f_0 = f : A \to B,$$
$$f_1 = f + \mathcal{P}_!(f) : A + \mathcal{P}(A) \to B + \mathcal{P}(B),$$
$$\vdots$$
$$f_{n+1} = f + \mathcal{P}_!(f_n) : A + \mathcal{P}(V_n(A)) \to B + \mathcal{P}(V_n(B)).$$

Since all the squares clearly commute, we have a cocone on the diagram of $V_n(A)$'s with vertex $V_\omega(B)$, and there is thus a unique $f_\omega : V_\omega(A) \to V_\omega(B)$ that completes the diagram.

Thus, we see that the cumulative hierarchy is functorial.

Example 5.33. $\omega CPOs$. An ωCPO is a poset that is "ω-cocomplete," meaning it has all colimits of type $\omega = (\mathbb{N}, \leq)$. Specifically, a poset D is an ωCPO if for every diagram $d : \omega \to D$, that is, every chain of elements of D,

$$d_0 \leq d_1 \leq d_2 \leq \cdots$$

we have a colimit $d_\omega = \varinjlim d_n$. This is an element of D such that

1. $d_n \leq d_\omega$ for all $n \in \omega$,
2. for all $x \in D$, if $d_n \leq x$ for all $n \in \omega$, then also $d_\omega \leq x$.

A monotone map of ωCPOs

$$h : D \to E$$

is called *continuous* if it preserves colimits of type ω, that is,

$$h(\varinjlim d_n) = \varinjlim h(d_n).$$

An application of these notions is the following.

Proposition 5.34. *If D is an ωCPO with initial element 0 and*

$$h : D \to D$$

is continuous, then h has a fixed point

$$h(x) = x$$

which, moreover, is least among all fixed points.

Proof. We use "Newton's method," which can be used, for example, to find fixed points of monotone, continuous functions $f : [0,1] \to [0,1]$. Consider the sequence $d : \omega \to D$, defined by

$$d_0 = 0$$

$$d_{n+1} = h(d_n)$$

Since $0 \leq d_0$, repeated application of h gives $d_n \leq d_{n+1}$. Now take the colimit $d_\omega = \varinjlim_{n \in \omega} d_n$. Then

$$h(d_\omega) = h(\varinjlim_{n \in \omega} d_n)$$

$$= \varinjlim_{n \in \omega} h(d_n)$$

$$= \varinjlim_{n \in \omega} d_{n+1}$$

$$= d_\omega.$$

The last step follows because the first term $d_0 = 0$ of the sequence is trivial.
Moreover, if x is also a fixed point, $h(x) = x$, then we have

$$d_0 = 0 \leq x$$

$$d_1 = h(0) \leq h(x) = x$$

$$\vdots$$

$$d_{n+1} = h(d_n) \leq h(x) = x.$$

So also $d_\omega \leq x$, since d_ω is the colimit. $\qquad\qquad\qquad\square$

Finally, here is an example of how (co)limits depend on the ambient category. We consider colimits *of* posets and ωCPOs, rather than *in* them.

Let us define the finite ωCPOs

$$\omega_n = \{k \leq n \mid k \in \omega\}$$

then we have continuous inclusion maps:

$$\omega_0 \to \omega_1 \to \omega_2 \to \cdots$$

In **Pos**, the colimit exists, and is ω, as can be easily checked. *But* ω itself is not ω-complete. Indeed, the sequence

$$0 \leq 1 \leq 2 \leq \cdots$$

has no colimit. Therefore, the colimit of the ω_n in the category of ωCPOs, if it exists, must be something else. In fact, it is $\omega + 1$.

$$0 \leq 1 \leq 2 \leq \cdots \leq \omega$$

For then any bounded sequence has a colimit in the bounded part, and any unbounded one has ω as colimit. The moral is that even ω-colimits are not always created in **Sets**, and indeed the colimit is sensitive to the ambient category in which it is taken.

5.7 Exercises

1. Show that a pullback of arrows

$$
\begin{array}{ccc}
A \times_X B & \xrightarrow{p_2} & B \\
{\scriptstyle p_1}\downarrow & & \downarrow{\scriptstyle g} \\
A & \xrightarrow{f} & X
\end{array}
$$

in a category **C** is the same thing as their product in the slice category **C**/X.

2. Let **C** be a category with pullbacks.

   (a) Show that an arrow $m : M \to X$ in **C** is monic if and only if the diagram below is a pullback.

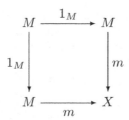

Thus, as an object in $\mathbf{C}/X$, m is monic iff $m \times m \cong m$.

(b) Show that the pullback along an arrow $f : Y \to X$ of a pullback square over X,

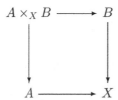

is again a pullback square over Y. (Hint: draw a cube and use the two-pullbacks lemma.) Conclude that the pullback functor f^* preserves products.

(c) Conclude from the foregoing that in a pullback square

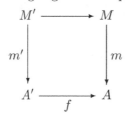

if m is monic, then so is m'.

3. Show directly that in any category, given a pullback square

$$
\begin{array}{ccc}
M' & \longrightarrow & M \\
\downarrow{\scriptstyle m'} & & \downarrow{\scriptstyle m} \\
A' & \xrightarrow{\ f\ } & A
\end{array}
$$

if m is monic, then so is m'.

4. For any object A in a category $\mathbf{C}$ and any subobjects $M, N \in \mathrm{Sub}_{\mathbf{C}}(A)$, show $M \subseteq N$ iff for every generalized element $z : Z \to A$ (arbitrary arrow with codomain A):

$$
z \in_A M \ \text{ implies } \ z \in_A N.
$$

5. Use the foregoing to show that for any object A in a category $\mathbf{C}$ and any subobjects $M, N \in \mathrm{Sub}_{\mathbf{C}}(A)$, show $M = N$ iff for every generalized element $z : Z \to A$ (arbitrary arrow with codomain A):

$$z \in_A M \quad \text{iff} \quad z \in_A N.$$

6. (Equalizers by pullbacks and products) Show that a category with pullbacks and products has equalizers as follows: given arrows $f, g : A \to B$, take the pullback indicated below, where $\Delta = \langle 1_B, 1_B \rangle$:

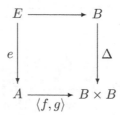

Show that $e : E \to A$ is the equalizer of f and g.

7. Let $\mathbf{C}$ be a locally small category with all small limits, and $D : \mathbf{J} \to \mathbf{C}$ any diagram in $\mathbf{C}$. Show that for any object $C \in \mathbf{C}$, the representable functor

$$\mathrm{Hom}_{\mathbf{C}}(C, -) : \mathbf{C} \to \mathbf{Sets}$$

preserves the limit of D.

8. (Partial maps) For any category $\mathbf{C}$ with pullbacks, define the category $\mathbf{Par}(\mathbf{C})$ of partial maps in $\mathbf{C}$ as follows: the objects are the same as those of $\mathbf{C}$, but an arrow $f : A \to B$ is a pair $(|f|, U_f)$, where $U_f \rightarrowtail A$ is a subobject and $|f| : U_f \to B$ is a suitable equivalence class of arrows, as indicated in the diagram:

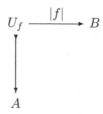

Composition of $(|f|, U_f) : A \to B$ and $(|g|, U_g) : B \to C$ is given by taking a pullback and then composing to get $(|g \circ f|, |f|^*(U_g))$, as suggested by the following diagram:

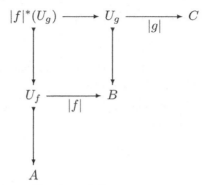

Verify that this really does define a category, and show that there is a functor,

$$\mathbf{C} \to \mathbf{Par}(\mathbf{C})$$

which is the identity on objects.

9. Suppose the category $\mathbf{C}$ has limits of type $\mathbf{J}$, for some index category $\mathbf{J}$. For diagrams F and G of type $\mathbf{J}$ in $\mathbf{C}$, a morphism of diagrams $\theta : F \to G$ consists of arrows $\theta_i : Fi \to Gi$ for each $i \in \mathbf{J}$ such that for each $\alpha : i \to j$ in $\mathbf{J}$, one has $\theta_j F(\alpha) = G(\alpha)\theta_i$ (a commutative square). This makes $\mathbf{Diagrams}(\mathbf{J}, \mathbf{C})$ into a category (check this).

Show that taking the vertex-objects of limiting cones determines a functor:

$$\varprojlim_{\mathbf{J}} : \mathbf{Diagrams}(J, \mathbf{C}) \to \mathbf{C}$$

Infer that for any set I, there is a product functor,

$$\prod_{i \in I} : \mathbf{Sets}^I \to \mathbf{Sets}$$

for I-indexed families of sets $(A_i)_{i \in I}$.

10. (Pushouts)

   (a) Dualize the definition of a pullback to define the "copullback" (usually called the "pushout") of two arrows with common domain.

   (b) Indicate how to construct pushouts using coproducts and coequalizers (proof "by duality").

11. Let $R \subseteq X \times X$ be an equivalence relation on a set X, with quotient $q : X \twoheadrightarrow Q$. Show that the following is an equalizer:

$$\mathcal{P}Q \xrightarrow{\mathcal{P}q} \mathcal{P}X \underset{\mathcal{P}r_2}{\overset{\mathcal{P}r_1}{\rightrightarrows}} \mathcal{P}R,$$

where $r_1, r_2 : R \rightrightarrows X$ are the two projections of $R \subseteq X$, and $\mathcal{P}$ is the (contravariant) powerset functor. (Hint: $\mathcal{P}X \cong 2^X$.)

12. Consider the sequence of posets $[0] \to [1] \to [2] \to \ldots$, where

$$[n] = \{0 \le \cdots \le n\},$$

and the arrows $[n] \to [n+1]$ are the evident inclusions. Determine the limit and colimit posets of this sequence.

13. Consider sequences of monoids,

$$M_0 \to M_1 \to M_2 \to \ldots$$
$$N_0 \leftarrow N_1 \leftarrow N_2 \leftarrow \ldots$$

and the following limits and colimits, constructed in the category of monoids:

$$\varinjlim_n M_n, \quad \varprojlim_n M_n, \quad \varinjlim_n N_n, \quad \varprojlim_n N_n.$$

(a) Suppose all M_n and N_n are abelian groups. Determine whether each of the four (co)limits $\varinjlim_n M_n$ etc. is also an abelian group.

(b) Suppose all M_n and N_n are finite groups. Determine whether each of the four (co)limits $\varinjlim_n M_n$ etc. has the following property: for every element x, there is a number k such that $x^k = 1$ (the least such k is called the *order* of x).

6

EXPONENTIALS

We have now managed to unify most of the universal mapping properties that we have seen so far with the notion of limits (or colimits). Of course, the free algebras are an exception to this. In fact, it turns out that there is a common source of such universal mapping properties, but it lies somewhat deeper, in the notion of *adjoints*, which unify free algebras, limits, and other universals of various kinds.

Next we are going to look at one more elementary universal structure, which is also an example of a universal that is not a limit. This important structure is called an "exponential," and it can be thought of as a categorical notion of a "function space." As we shall see it subsumes much more than just that, however.

6.1 Exponential in a category

Let us start by considering a function of sets,

$$f(x, y) : A \times B \to C$$

written using variables x over A and y over B. If we now hold $a \in A$ fixed, we have a function

$$f(a, y) : B \to C$$

and thus an element

$$f(a, y) \in C^B$$

of the set of all such functions.

Letting a vary over A then gives a map, which I write like this

$$\tilde{f} : A \to C^B$$

defined by $a \mapsto f(a, y)$.

The map $\tilde{f} : A \to C^B$ takes the "parameter" a to the function $f(a, y) : B \to C$. It is uniquely determined by the equation

$$\tilde{f}(a)(b) = f(a, b).$$

Indeed, *any* map

$$\phi : A \to C^B$$

is uniquely of the form

$$\phi = \tilde{f}$$

for some $f : A \times B \to C$. For we can set

$$f(a,b) := \phi(a)(b).$$

What this means, in sum, is that we have an isomorphism of Hom-sets:

$$\mathrm{Hom}_{\mathbf{Sets}}(A \times B, C) \cong \mathrm{Hom}_{\mathbf{Sets}}(A, C^B)$$

That is, there is a bijective correspondence between functions of the form $f : A \times B \to C$ and those of the form $\tilde{f} : A \to C^B$, which we can display schematically thus

$$\frac{f : A \times B \to C}{\tilde{f} : A \to C^B}$$

This bijection is mediated by a certain operation of *evaluation*, which we have indicated in the foregoing by using variables. In order to generalize the indicated bijection to other categories, we are going to need to make this evaluation operation explicit, too.

In **Sets**, it is the function

$$\mathrm{eval} : C^B \times B \to C$$

defined by $(g,b) \mapsto g(b)$, that is,

$$\mathrm{eval}(g, b) = g(b).$$

This evaluation function has the following UMP: given any set A and any function

$$f : A \times B \to C$$

there is a unique function

$$\tilde{f} : A \to C^B$$

such that $\mathrm{eval} \circ (\tilde{f} \times 1_B) = f$. That is,

$$\mathrm{eval}(\tilde{f}(a), b) = f(a, b). \tag{6.1}$$

Here is the diagram:

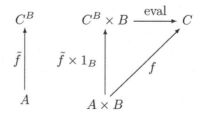

You can read the equation (6.1) off from this diagram by taking a pair of elements $(a, b) \in A \times B$ and chasing them around both ways, using the fact that $(\tilde{f} \times 1_B)(a, b) = (\tilde{f}(a), b)$.

Now, the property just stated of the set C^B and the evaluation function eval : $C^B \times B \to C$ is one that makes sense in any category having binary products. It says that evaluation is "the universal map into C from a product with B." Precisely, we have the following:

Definition 6.1. Let the category **C** have binary products. An *exponential* of objects B and C consists of an object

$$C^B$$

and an arrow

$$\epsilon : C^B \times B \to C$$

such that, for any object A and arrow

$$f : A \times B \to C$$

there is a unique arrow

$$\tilde{f} : A \to C^B$$

such that

$$\epsilon \circ (\tilde{f} \times 1_B) = f$$

all as in the diagram

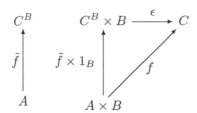

Here is some terminology:

- $\epsilon : C^B \times B \to C$ is called *evaluation*.
- $\tilde{f} : A \to C^B$ is called the (exponential) *transpose* of f.
- Given any arrow

$$g : A \to C^B$$

  we write

$$\bar{g} := \epsilon \circ (g \times 1_B) : A \times B \to C$$

and also call $\bar{g}$ the *transpose* of g. By the uniqueness clause of the definition, we then have

$$\tilde{\bar{g}} = g$$

and for any $f : A \times B \to C$,

$$\bar{\tilde{f}} = f.$$

Briefly, transposition of transposition is the identity.

Thus in sum, the transposition operation

$$(f : A \times B \to C) \longmapsto (\tilde{f} : A \to C^B)$$

provides an inverse to the induced operation

$$(g : A \to C^B) \longmapsto (\bar{g} = \epsilon \circ (g \times 1_B) : A \times B \to C),$$

yielding the desired isomorphism,

$$\mathrm{Hom}_{\mathbf{C}}(A \times B, C) \cong \mathrm{Hom}_{\mathbf{C}}(A, C^B).$$

6.2   Cartesian closed categories

Definition 6.2. A category is called *cartesian closed*, if it has all finite products and exponentials.

Example 6.3. We already have **Sets** as one example, but note that also $\mathbf{Sets}_{fin}$ is cartesian closed, since for finite sets M, N, the set of functions N^M has cardinality

$$|N^M| = |N|^{|M|}$$

and so is also finite.

Example 6.4. Recall that the category **Pos** of posets has as arrows $f : P \to Q$ the monotone functions, $p \le p'$ implies $fp \le fp'$. Given posets P and Q, the poset $P \times Q$ has pairs (p, q) as elements, and is partially ordered by

$$(p, q) \le (p', q') \quad \text{iff} \quad p \le p' \text{ and } q \le q'.$$

Thus, the evident projections

$$P \xleftarrow{\ \pi_1\ } P \times Q \xrightarrow{\ \pi_2\ } Q$$

are monotone, as is the pairing

$$\langle f, g \rangle : X \to P \times Q$$

if $f : X \to P$ and $g : X \to Q$ are monotone.

For the exponential Q^P, we take the set of monotone functions,

$$Q^P = \{f : P \to Q \mid f \text{ monotone }\}$$

ordered *pointwise*, that is,

$$f \le g \quad \text{iff} \quad fp \le gp \text{ for all } p \in P.$$

The evaluation

$$\epsilon : Q^P \times P \to Q$$

and transposition

$$\tilde{f} : X \to Q^P$$

of a given arrow

$$f : X \times P \to Q$$

are the usual ones of the underlying functions. Thus, we need only show that these are monotone.

To that end, given $(f, p) \le (f', p')$ in $Q^P \times P$, we have

$$\begin{aligned} \epsilon(f, p) &= f(p) \\ &\le f(p') \\ &\le f'(p') \\ &= \epsilon(f', p') \end{aligned}$$

so ϵ is monotone. Now take $f : X \times P \to Q$ monotone and let $x \le x'$. We need to show

$$\tilde{f}(x) \le \tilde{f}(x') \quad \text{in } Q^P$$

which means

$$\tilde{f}(x)(p) \le \tilde{f}(x')(p) \quad \text{for all } p \in P.$$

But $\tilde{f}(x)(p) = f(x, p) \le f(x', p) = \tilde{f}(x')(p)$.

Example 6.5. Now let us consider what happens if we restrict to the category of ωCPOs (see example 5.33). Given two ωCPOs P and Q, we take as an exponential the subset,

$$Q^P = \{f : P \to Q \mid f \text{ monotone and } \omega\text{-continuous}\}.$$

Then take evaluation $\epsilon : Q^P \times P \to Q$ and transposition as before, for functions. Then, since we know that the required equations are satisfied, we just need to check the following:

• Q^P is an ωCPO

- ϵ is ω-continuous
- $\tilde{f}$ is ω-continuous if f is so

We leave this as an exercise!

Example 6.6. An example of a somewhat different sort is provided by the category **Graphs** of graphs and their homomorphisms. Recall that a graph G consists of a pair of sets G_e and G_v—the edges and vertices—and a pair of functions,

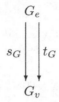

called the source and target maps. A homomorphism of graphs $h : G \to H$ is a mapping of edges to edges and vertices to vertices, preserving sources and targets, that is, is a pair of maps $h_v : G_v \to H_v$ and $h_e : G_e \to H_e$, making the two obvious squares commute.

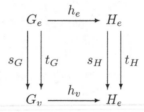

The product $G \times H$ of two graphs G and H, like the product of categories, has as vertices the pairs (g, h) of vertices $g \in G$ and $h \in H$, and similarly the edges are pairs of edges (u, v) with u an edge in G and v and edge in H. The source and target operations are, then, "pointwise": $s(u, v) = (s(u), s(v))$, etc.

$$
\begin{array}{c}
G_e \times H_e \\
\\
s_G \times s_H \quad\bigg\downarrow\bigg\downarrow\quad t_G \times t_H \\
\\
G_v \times H_v
\end{array}
$$

Now, the exponential graph H^G has as vertices the (arbitrary!) maps of vertices $\varphi : G_v \to H_v$. An edge θ from φ to another vertex $\psi : G_v \to H_v$ is a family of edges (θ_e) in H, one for each edge $e \in G$, such that $s(\theta_e) = \varphi(s(e))$ and $t(\theta_e) = \psi(t(e))$. In other words, θ is a map $\theta : G_e \to H_e$ making the following commute:

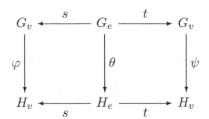

Imagining G as a certain configuration of edges and vertices, and the maps φ and ψ as two different "pictures" or "images" of the vertices of G in H, the edge $\theta : \varphi \to \psi$ appears as a family of edges in H, labeled by the edges of G, each connecting the source vertex in φ to the corresponding target one in ψ. (The reader should draw a diagram at this point.) The evaluation homomorphism $\epsilon : H^G \times G \to H$ takes a vertex (φ, g) to the vertex $\psi(y)$, and an edge (θ, e) to the edge θ_e. The transpose of a graph homomorphism $f : F \times G \to H$ is the homomorphism $\tilde{f} : F \to H^G$ taking a vertex $a \in F$ to the mapping on vertices $f(a, -) : G_v \to H_v$, and an edge $c : a \to b$ in F to the mapping of edges $f(c, -) : G_e \to H_e$.

We leave the verification of this cartesian closed structure as an exercise for the reader.

Next, we derive some of the basic facts about exponentials and cartesian closed categories. First, let us ask, what is the transpose of evaluation?

$$\epsilon : B^A \times A \to B$$

It must be an arrow $\tilde{\epsilon} : B^A \to B^A$ such that

$$\epsilon(\tilde{\epsilon} \times 1_A) = \epsilon$$

that is, making the following diagram commute:

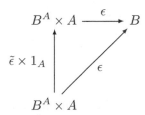

Since $1_{B^A} \times 1_A = 1_{(B^A \times A)}$ clearly has this property, we must have

$$\tilde{\epsilon} = 1_{B^A}$$

and so we also know that $\epsilon = \overline{(1_{B^A})}$.

Now let us show that the operation $X \mapsto X^A$ on a CCC is *functorial*.

Proposition 6.7. *In any cartesian closed category* **C**, *exponentiation by a fixed object A is a functor,*

$$(-)^A : \mathbf{C} \to \mathbf{C}.$$

Toward the proof, consider first the case of sets. Given some function

$$\beta : B \to C,$$

we put

$$\beta^A : B^A \to C^A$$

defined by

$$f \mapsto \beta \circ f.$$

That is,

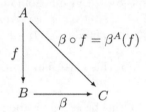

This assignment is functorial, because for any $\alpha : C \to D$

$$(\alpha \circ \beta)^A(f) = \alpha \circ \beta \circ f$$
$$= \alpha \circ \beta^A(f)$$
$$= \alpha^A \circ \beta^A(f).$$

Whence $(\alpha \circ \beta)^A = \alpha^A \circ \beta^A$. Also,

$$(1_B)^A(f) = 1_B \circ f$$
$$= f$$
$$= 1_{B^A}(f).$$

So $(1_B)^A = 1_{B^A}$. Thus, $(-)^A$ is indeed a functor; of course, it is just the representable functor $\mathrm{Hom}(A, -)$ that we have already considered.

In a general CCC then, given $\beta : B \to C$, we define

$$\beta^A : B^A \to C^A$$

by

$$\beta^A := \widetilde{(\beta \circ \epsilon)}.$$

That is, we take the transpose of the composite

$$B^A \times A \xrightarrow{\epsilon} B \xrightarrow{\beta} C$$

giving

$$\beta^A : B^A \to C^A.$$

It is easier to see in the form

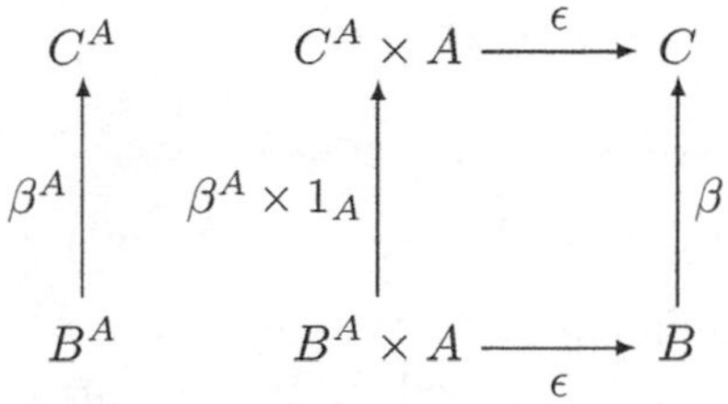

Now, clearly,

$$(1_B)^A = 1_{B^A} : B^A \to B^A$$

by examining

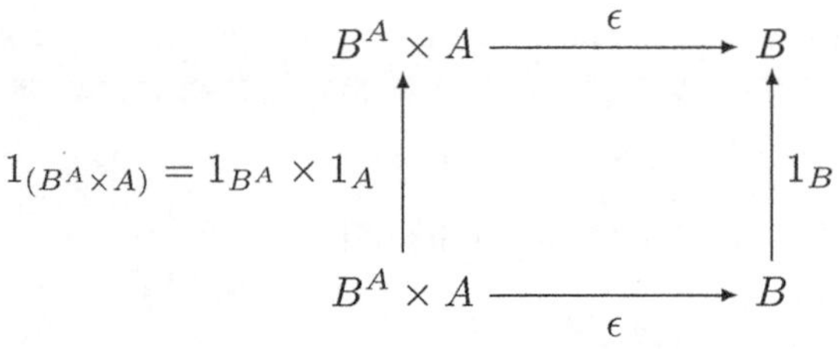

Quite similarly, given

$$B \xrightarrow{\beta} C \xrightarrow{\gamma} D$$

we have

$$\gamma^A \circ \beta^A = (\gamma \circ \beta)^A.$$

This follows from considering the commutative diagram:

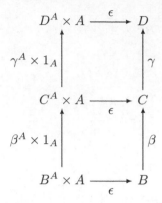

We use the fact that

$$(\gamma^A \times 1_A) \circ (\beta^A \times 1_A) = ((\gamma^A \circ \beta^A) \times 1_A).$$

The result follows by the uniqueness of transposes.

There is also another distinguished "universal" arrow; rather than transposing $1_{B^A} : B^A \to B^A$, we can transpose the identity $1_{A \times B} : A \times B \to A \times B$, to get

$$\tilde{1}_{A \times B} : A \to (A \times B)^B.$$

In **Sets**, it has the values $\tilde{1}_{A \times B}(a)(b) = (a, b)$. Let us denote this map by $\eta = \tilde{1}_{A \times B}$, so that

$$\eta(a)(b) = (a, b).$$

The map η lets us compute $\tilde{f}$ from the functor $-^A$. Indeed, given $f : Z \times A \to B$, take

$$f^A : (Z \times A)^A \to B^A$$

and precompose with $\eta : Z \to (Z \times A)^A$, as indicated in

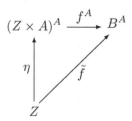

This gives the useful equation

$$\tilde{f} = f^A \circ \eta$$

which the reader should prove.

6.3   Heyting algebras

Any Boolean algebra B, regarded as a poset category, has finite products 1 and $a \wedge b$. We can also define the exponential in B by

$$b^a = (\neg a \vee b)$$

which we also write $a \Rightarrow b$. The evaluation arrow is

$$(a \Rightarrow b) \wedge a \leq b.$$

This always holds since

$$(\neg a \vee b) \wedge a = (\neg a \wedge a) \vee (b \wedge a) = 0 \vee (b \wedge a) = b \wedge a \leq b.$$

To show that $a \Rightarrow b$ is indeed an exponential in B, we just need to verify that if $a \wedge b < c$ then $a < b \Rightarrow c$, that is, transposition. But if $a \wedge b \leq c$, then

$$\neg b \vee (a \wedge b) \leq \neg b \vee c = b \Rightarrow c.$$

But we also have

$$a \leq \neg b \vee a \leq (\neg b \vee a) \wedge (\neg b \vee b) = \neg b \vee (a \wedge b).$$

This example suggests generalizing the notion of a Boolean algebra to that of a cartesian closed poset. Indeed, consider first the following stronger notion.

Definition 6.8. A *Heyting algebra* is a poset with

1. Finite meets: 1 and $p \wedge q$,
2. Finite joins: 0 and $p \vee q$,
3. Exponentials: for each a, b, an element $a \Rightarrow b$ such that

$$a \wedge b \leq c \quad \text{iff} \quad a \leq b \Rightarrow c.$$

The stated condition on exponentials $a \Rightarrow b$ is equivalent to the UMP in the case of posets. Indeed, given the condition, the transpose of $a \wedge b \leq c$ is $a \leq b \Rightarrow c$ and the evaluation $(a \Rightarrow b) \wedge a \leq b$ follows immediately from $a \Rightarrow b \leq a \Rightarrow b$ (the converse is just as simple).

First, observe that every Heyting algebra is a *distributive lattice*, that is, for any a, b, c, one has

$$(a \vee b) \wedge c = (a \wedge c) \vee (b \wedge c).$$

Indeed, we have

$$(a \vee b) \wedge c \leq z \text{ iff } a \vee b \leq c \Rightarrow z$$
$$\text{iff } a \leq c \Rightarrow z \text{ and } b \leq c \Rightarrow z$$
$$\text{iff } a \wedge c \leq z \text{ and } b \wedge c \leq z$$
$$\text{iff } (a \wedge c) \vee (b \wedge c) \leq z.$$

Now pick $z = (a \vee b) \wedge c$ and read the equivalences downward to get one direction, then do the same with $z = (a \wedge c) \vee (b \wedge c)$ and reading the equivalences upward to get the other direction.

Remark 6.9. The foregoing distributivity is actually a special case of the more general fact that in a cartesian closed category with coproducts, the products necessarily distribute over the coproducts,

$$(A + B) \times C \cong (A \times C) + (B \times C).$$

Although we could prove this now directly, a much more elegant proof (generalizing the one above for the poset case) will be available to us once we have access to the Yoneda lemma. For this reason, we defer the proof of distributivity to 8.6.

One may well wonder whether all distributive lattices are Heyting algebras. The answer is in general, no; but certain ones always are.

Definition 6.10. A poset is (*co*)*complete* if it is so as a category, thus if it has all set-indexed meets $\bigwedge_{i \in I} a_i$ (resp. joins $\bigvee_{i \in I} a_i$). A lattice, Heyting algebra, Boolean algebra, etc. is called *complete* if it is so as a poset. For lattices, completeness and cocompleteness are equivalent (exercise!).

Proposition 6.11. *A complete lattice is a Heyting algebra iff it satisfies the infinite distributive law*

$$a \wedge \left(\bigvee_i b_i \right) = \bigvee_i (a \wedge b_i).$$

Proof. One shows that Heyting algebra implies distributivity just as in the finite case. To show that the infinite distributive law implies Heyting algebra, set

$$a \Rightarrow b = \bigvee_{x \wedge a \leq b} x.$$

Then, if

$$y \wedge a \leq b$$

then $y \leq \bigvee_{x \wedge a \leq b} x = a \Rightarrow b$. And conversely, if $y \leq a \Rightarrow b$, then $y \wedge a \leq (\bigvee_{x \wedge a \leq b} x) \wedge a = \bigvee_{x \wedge a \leq b} (x \wedge a) \leq \bigvee b = b$. $\square$

Example 6.12. For any set A, the powerset $P(A)$ is a complete Heyting algebra with unions and intersections as joins and meets, since it satisfies the infinite distributive law. More generally, the lattice of open sets of a topological space is also a Heyting algebra, since the open sets are closed under finite intersections and arbitrary unions.

Of course, every Boolean algebra is a Heyting algebra with $a \Rightarrow b = \neg a \vee b$, as we already showed. But in general, a Heyting algebra is not Boolean. Indeed, we can define a proposed negation by

$$\neg a = a \Rightarrow 0$$

as must be the case, since in a Boolean algebra $\neg a = \neg a \vee 0 = a \Rightarrow 0$. Then $a \leq \neg\neg a$ since $a \wedge (a \Rightarrow 0) \leq 0$. But, conversely, $\neg\neg a \leq a$ need not hold in a Heyting algebra. Indeed, in a topological space X, the negation $\neg U$ of an open subset U is the *interior* of the complement $X - U$. Thus, for example, in the real interval $[0, 1]$, we have $\neg\neg(0, 1) = [0, 1]$.

Moreover, the law,

$$1 \leq a \vee \neg a$$

also need not hold in general. In fact, the concept of a Heyting algebra is the algebraic equivalent of the *intuitionistic* propositional calculus, in the same sense that Boolean algebras are an algebraic formulation of the *classical* propositional calculus.

6.4   Propositional calculus

In order to make the connection between Heyting algebras and propositional calculus more rigorous, let us first give a specific system of rules for the intuitionistic propositional calculus (IPC). This we do in terms of entailments $p \vdash q$ between formulas p and q:

1. $\vdash$ is reflexive and transitive
2. $p \vdash \top$
3. $\perp \vdash p$
4. $p \vdash q$ and $p \vdash r$ iff $p \vdash q \wedge r$
5. $p \vdash r$ and $q \vdash r$ iff $p \vee q \vdash r$
6. $p \wedge q \vdash r$ iff $p \vdash q \Rightarrow r$

This is a complete system for IPC, equivalent to the more standard presentations the reader may have seen. To compare with one perhaps more familiar presentation, note first that we have an "evaluation" entailment by reflexivity and (6):

$$p \Rightarrow q \vdash p \Rightarrow q$$

$$(p \Rightarrow q) \wedge p \vdash q$$

We therefore have the rule of "modus ponens" by (4) and transitivity:

$$\top \vdash p \Rightarrow q \quad \text{and} \quad \top \vdash p$$

$$\top \vdash (p \Rightarrow q) \wedge p$$

$$\top \vdash q$$

Moreover, by (4) there are "projections":

$$p \wedge q \vdash p \wedge q$$
$$p \wedge q \vdash p \quad (\text{resp. } q)$$

from which it follows that $p \dashv\vdash \top \wedge p$. Thus, we get one of the usual axioms for products:

$$p \wedge q \vdash p$$
$$\top \wedge (p \wedge q) \vdash p$$
$$\top \vdash (p \wedge q) \Rightarrow p$$

Now let us derive the usual axioms for $\Rightarrow$, namely,

1. $p \Rightarrow p$,
2. $p \Rightarrow (q \Rightarrow p)$,
3. $(p \Rightarrow (q \Rightarrow r)) \Rightarrow ((p \Rightarrow q) \Rightarrow (p \Rightarrow r))$.

The first two are almost immediate:

$$p \vdash p$$
$$\top \wedge p \vdash p$$
$$\top \vdash p \Rightarrow p$$

$$p \wedge q \vdash p$$
$$p \vdash q \Rightarrow p$$
$$\top \wedge p \vdash (q \Rightarrow p)$$
$$\top \vdash p \Rightarrow (q \Rightarrow p)$$

For the third one, we use the fact that $\Rightarrow$ distributes over $\wedge$ on the right:

$$a \Rightarrow (b \wedge c) \dashv\vdash (a \Rightarrow b) \wedge (a \Rightarrow c)$$

This is a special case of the exercise:

$$(B \times C)^A \cong B^A \times C^A$$

We also use the following simple fact, which will be recognized as a special case of proposition 6.7:

$$a \vdash b \quad \text{implies} \quad p \Rightarrow a \vdash p \Rightarrow b \tag{6.2}$$

Then we have

$$(q \Rightarrow r) \wedge q \vdash r$$

$$p \Rightarrow ((q \Rightarrow r) \wedge q) \vdash p \Rightarrow r$$

$$(p \Rightarrow (q \Rightarrow r)) \wedge (p \Rightarrow q) \vdash p \Rightarrow r \qquad \text{by (6.3)}$$

$$(p \Rightarrow (q \Rightarrow r)) \vdash (p \Rightarrow q) \Rightarrow (p \Rightarrow r)$$

$$\top \vdash (p \Rightarrow (q \Rightarrow r)) \Rightarrow ((p \Rightarrow q) \Rightarrow (p \Rightarrow r)).$$

The "positive" fragment of IPC, involving only the logical operations

$$\top, \wedge, \Rightarrow$$

corresponds to the notion of a cartesian closed poset. We then add $\bot$ and disjunction $p \vee q$ on the logical side and finite joins on the algebraic side to arrive at a correspondence between IPC and Heyting algebras. The exact correspondence is given by mutually inverse constructions between Heyting algebras and IPCs. We briefly indicate one direction of this correspondence, leaving the other one to the reader's ingenuity.

Given any IPSs $\mathcal{L}$, consisting of propositional formulas $p, q, r, \dots$ over some set of variables $x, y, z, \dots$ together with the rules of inference stated above, and perhaps some distinguished formulas $a, b, c, \dots$ as axioms, one constructs from $\mathcal{L}$ a Heyting algebra $\mathrm{HA}(\mathcal{L})$, called the *Lindenbaum–Tarski algebra*, consisting of equivalence classes $[p]$ of formulas p, where

$$[p] = [q] \quad \text{iff} \quad p \dashv\vdash q \qquad (6.3)$$

The ordering in $\mathrm{HA}(\mathcal{L})$ is given by

$$[p] \leq [q] \quad \text{iff} \quad p \vdash q \qquad (6.4)$$

This is clearly well defined on equivalence classes, in the sense that if $p \vdash q$ and $[p] = [p']$ then $p' \vdash q$, and similarly for q. The operations in $\mathrm{HA}(\mathcal{L})$ are then induced in the expected way by the logical operations:

$$1 = [\top]$$

$$0 = [\bot]$$

$$[p] \wedge [q] = [p \wedge q]$$

$$[p] \vee [q] = [p \vee q]$$

$$[p] \Rightarrow [q] = [p \Rightarrow q]$$

Again, these operations are easily seen to be well defined on equivalence classes, and they satisfy the laws for a Heyting algebra because the logical rules evidently imply them.

Lemma 6.13. *Observe that, by* (6.3), *the Heyting algebra* $\mathrm{HA}(\mathcal{L})$ *has the property that a formula p is provable* $\top \vdash p$ *if and only if* $[p] = 1$.

Now define an *interpretation* M of $\mathcal{L}$ in a Heyting algebra H to be an assignment of the basic propositional variables $x, y, z, \ldots$ to elements of H, which we shall write as $[\![x]\!], [\![y]\!], [\![z]\!], \ldots$. An interpretation then extends to all formulas by recursion in the evident way, that is, $[\![p \wedge q]\!] = [\![p]\!] \wedge [\![q]\!]$, etc. An interpretation is called a *model* of $\mathcal{L}$ if for every theorem $\top \vdash p$, one has $[\![p]\!] = 1$. Observe that there is a canonical interpretation of $\mathcal{L}$ in $\mathrm{HA}(\mathcal{L})$ given by $[\![x]\!] = [x]$. One shows easily by induction that, for any formula p, moreover, $[\![p]\!] = [p]$. Now lemma 6.13 tells us that this interpretation is in fact a model of $\mathcal{L}$ and that, moreover, it is "generic," in the sense that it validates *only* the provable formulas. We therefore have the following logical *completeness theorem for IPC*.

Proposition 6.14. *The intuitionistic propositional calculus is complete with respect to models in Heyting algebras.*

Proof. Suppose a formula p is true in all models in all Heyting algebras. Then in particular, it is so in $\mathrm{HA}(\mathcal{L})$. Thus, $1 = [\![p]\!] = [p]$ in $\mathrm{HA}(\mathcal{L})$, and so $\top \vdash p$. $\square$

In sum, then, a particular instance $\mathcal{L}$ of IPC can be regarded as a way of specifying (and reasoning about) a particular Heyting algebra $\mathrm{HA}(\mathcal{L})$. Indeed, it is essentially a presentation by generators and relations, in just the way that we have already seen for other algebraic objects like monoids. The Heyting algebra $\mathrm{HA}(\mathcal{L})$ even has a UMP with respect to $\mathcal{L}$ that is entirely analogous to the UMP of a finitely presented monoid given by generators and relations. Specifically, if, for instance, $\mathcal{L}$ is generated by the two elements x, y subject to the single "axiom" $x \vee y \Rightarrow x \wedge y$, then in $\mathrm{HA}(\mathcal{L})$ the elements $[x]$ and $[y]$ satisfy $[x] \vee [y] \leq [x] \wedge [y]$ (which is of course equivalent to $([x] \vee [y] \Rightarrow [x] \wedge [y]) = 1$), and given any Heyting algebra A with two elements a and b satisfying $a \vee b \leq a \wedge b$, there is a unique Heyting homomorphism $h : \mathrm{HA}(\mathcal{L}) \to A$ with $h([x]) = a$ and $h([y]) = b$. In this sense, the Lindenbaum–Tarski Heyting algebra $\mathrm{HA}(\mathcal{L})$, being finitely presented by the generators and axioms of $\mathcal{L}$, can be said to contain a "universal model" of the theory determined by $\mathcal{L}$.

6.5 Equational definition of CCC

The following description of CCCs in terms of operations and equations on a category is often useful. The proof is entirely routine and left to the reader.

Proposition 6.15. *A category* **C** *is a CCC iff it has the following structure:*

- *A distinguished object* 1*, and for each object* C *there is given an arrow*

$$!_C : C \to 1$$

such that for each arrow $f : C \to 1$*,*

$$f = !_C.$$

- *For each pair of objects A, B, there is given an object $A \times B$ and arrows,*

$$p_1 : A \times B \to A \quad and \quad p_2 : A \times B \to B$$

and for each pair of arrows $f : Z \to A$ and $g : Z \to B$, there is given an arrow,

$$\langle f, g \rangle : Z \to A \times B$$

such that

$$p_1 \langle f, g \rangle = f$$
$$p_2 \langle f, g \rangle = g$$
$$\langle p_1 h, p_2 h \rangle = h \quad for \ all \ h : Z \to A \times B.$$

- *For each pair of objects A, B, there is given an object B^A and an arrow,*

$$\epsilon : B^A \times A \to B$$

and for each arrow $f : Z \times A \to B$, there is given an arrow

$$\tilde{f} : Z \to B^A$$

such that

$$\epsilon \circ (\tilde{f} \times 1_A) = f$$

and

$$(\widetilde{\epsilon \circ (g \times 1_A)}) = g$$

for all $g : Z \to B^A$. Here, and generally, for any $a : X \to A$ and $b : Y \to B$, we write

$$a \times b = \langle a \circ p_1, b \circ p_2 \rangle : X \times Y \to A \times B.$$

It is sometimes easier to check these equational conditions than to verify the corresponding UMPs. Section 6.6 provides an example of this sort.

6.6 λ-calculus

We have seen that the notions of a cartesian closed poset with finite joins (i.e., a Heyting algebra) and intuitionistic propositional calculus are essentially the same:

$$\mathrm{HA} \sim \mathrm{IPC}.$$

These are two different ways of describing one and the same structure; whereby, to be sure, the logical description contains some superfluous data in the choice of a particular presentation.

We now want to consider another, very similar, correspondence between systems of logic and categories, involving more general CCCs. Indeed, the fore-

going correspondence was the poset case of the following general one between
CCCs and λ-calculus:

$$\text{CCC} \sim \lambda\text{-calculus.}$$

These notions are also essentially equivalent, in a sense that we now sketch (a
more detailed treatment can be found in the book by Lambek and Scott). They
are two different ways of representing the same idea, namely that of a collection
of objects and arrows, with operations of *pairing, projection, application,* and
transposition (or "*currying*").

First, recall the notion of a (typed) λ-calculus from Chapter 2. It consists
of

- Types: $A \times B$, $A \to B, \ldots$ (and some basic types)
- Terms: $x, y, z, \ldots : A$ (variables for each type A)
  $a : A$, $b : B, \ldots$ (possibly some typed constants)

$$\langle a, b \rangle : A \times B \quad (a : A,\ b : B)$$

$$\text{fst}(c) : A \qquad (c : A \times B)$$

$$\text{snd}(c) : B \qquad (c : A \times B)$$

$$ca : B \qquad (c : A \to B,\ a : A)$$

$$\lambda x.b : A \to B \quad (x : A,\ b : B)$$

- Equations, including at least all instances of the following:

$$\text{fst}(\langle a, b \rangle) = a$$

$$\text{snd}(\langle a, b \rangle) = b$$

$$\langle \text{fst}(c), \text{snd}(c) \rangle = c$$

$$(\lambda x.b)a = b[a/x]$$

$$\lambda x.cx = c \quad (\text{no } x \text{ in } c)$$

Given a particular such λ-calculus $\mathcal{L}$, the associated *category of types* $\mathbf{C}(\mathcal{L})$
was then defined as follows:

- Objects: the types,
- Arrows $A \to B$: equivalence classes of closed terms $[c] : A \to B$, identified
  according to (renaming of bound variables and),

$$[a] = [b] \quad \text{iff } \mathcal{L} \vdash a = b \tag{6.5}$$

- Identities: $1_A = [\lambda x.x]$ (where $x : A$),
- Composition: $[c] \circ [b] = [\lambda x.c(bx)]$.

We have already seen that this is a well-defined category, and that it has
binary products. It is a simple matter to add a terminal object. Now let us use

the equational characterization of CCCs to show that it is cartesian closed. Given any objects A, B, we set $B^A = A \to B$, and as the evaluation arrow, we take (the equivalence class of),

$$\epsilon = \lambda z.\mathrm{fst}(z)\mathrm{snd}(z) : B^A \times A \to B \quad (z : Z).$$

Then for any arrow $f : Z \times A \to B$, we take as the transpose,

$$\tilde{f} = \lambda z \lambda x.f\langle z, x \rangle : Z \to B^A \quad (z : Z, \ x : A).$$

It is now a straightforward λ-calculus calculation to verify the two required equations, namely,

$$\epsilon \circ (\tilde{f} \times 1_A) = f,$$

$$(\epsilon \circ \widetilde{(g \times 1_A)}) = g.$$

In detail, for the first one recall that

$$\alpha \times \beta = \lambda w.\langle \alpha\mathrm{fst}(w), \beta\mathrm{snd}(w) \rangle.$$

So, we have

$$
\begin{aligned}
\epsilon \circ (\tilde{f} \times 1_A) &= (\lambda z.\mathrm{fst}(z)\mathrm{snd}(z)) \circ [(\lambda y \lambda x.f\langle y, x\rangle) \times \lambda u.u] \\
&= \lambda v.(\lambda z.\mathrm{fst}(z)\mathrm{snd}(z))[(\lambda y \lambda x.f\langle y, x\rangle) \times \lambda u.u]v \\
&= \lambda v.(\lambda z.\mathrm{fst}(z)\mathrm{snd}(z))[\lambda w.\langle(\lambda y \lambda x.f\langle y, x\rangle)\mathrm{fst}(w), (\lambda u.u)\mathrm{snd}(w)\rangle]v \\
&= \lambda v.(\lambda z.\mathrm{fst}(z)\mathrm{snd}(z))[\lambda w.\langle(\lambda x.f\langle\mathrm{fst}(w), x\rangle), \mathrm{snd}(w)\rangle]v \\
&= \lambda v.(\lambda z.\mathrm{fst}(z)\mathrm{snd}(z))[\langle(\lambda x.f\langle\mathrm{fst}(v), x\rangle), \mathrm{snd}(v)\rangle] \\
&= \lambda v.(\lambda x.f\langle\mathrm{fst}(v), x\rangle)\mathrm{snd}(v) \\
&= \lambda v.f\langle\mathrm{fst}(v), \mathrm{snd}(v)\rangle \\
&= \lambda v.fv \\
&= f.
\end{aligned}
$$

The second equation is proved similarly.

Let us call a set of basic types and terms, together with a set of equations between terms, a *theory* in the λ-calculus. Given such a theory $\mathcal{L}$, the cartesian closed category $\mathbf{C}(\mathcal{L})$ built from the λ-calculus over $\mathcal{L}$ is the CCC presented by the generators and relations stated by $\mathcal{L}$. Just as in the poset case of IPC and Heyting algebras, there is a logical completeness theorem that follows from this fact. To state it, we require the notion of a model of a theory $\mathcal{L}$ in the λ-calculus in an arbitrary cartesian closed category $\mathbf{C}$. We give only a brief sketch to give the reader the general idea.

Definition 6.16. A *model* of $\mathcal{L}$ in $\mathbf{C}$ is an assignment of the types and terms of $\mathcal{L}$ to objects and arrows of $\mathbf{C}$:

$$X \text{ basic type} \quad \leadsto \quad [\![X]\!] \text{ object}$$
$$b : A \to B \text{ basic term} \quad \leadsto \quad [\![b]\!] : [\![A]\!] \to [\![B]\!] \text{ arrow}$$

This assignment is then extended to all types and terms in such a way that the λ-calculus operations are taken to the corresponding CCC ones:

$$[\![A \times B]\!] = [\![A]\!] \times [\![B]\!]$$
$$[\![\langle f, g \rangle]\!] = \langle [\![f]\!], [\![g]\!] \rangle$$

etc.

Finally, it is required that all the equations of $\mathcal{L}$ are satisfied, in the sense that

$$\mathcal{L} \vdash a = b : A \to B \quad \text{implies} \quad [\![a]\!] = [\![b]\!] : [\![A]\!] \to [\![B]\!]. \tag{6.6}$$

This is what is sometimes called "denotational semantics" for the λ-calculus. It is essentially the conventional, set-theoretic semantics for first-order logic, but extended to higher types, restricted to equational theories, and generalized to CCCs.

For example, let $\mathcal{L}$ be the theory with one basic type X, two basic terms,

$$u : X$$
$$m : X \times X \to X$$

and the usual equations for associativity and units,

$$m\langle u, x \rangle = x$$
$$m\langle x, u \rangle = x$$
$$m\langle x, m\langle y, z \rangle \rangle = m\langle m\langle x, y \rangle, z \rangle.$$

Thus, $\mathcal{L}$ is just the usual equational theory of monoids. Then a model of $\mathcal{L}$ in a cartesian closed category $\mathbf{C}$ is nothing but a monoid in $\mathbf{C}$, that is, an object $M = [\![X]\!]$ equipped with a distinguished point

$$[\![u]\!] : 1 \to M$$

and a binary operation

$$[\![m]\!] : M \times M \to M$$

satisfying the unit and associativity laws.

Note that by (6.5) and (6.6), there is a model of $\mathcal{L}$ in $\mathbf{C}(\mathcal{L})$ with the property that $[\![a]\!] = [\![b]\!] : X \to Y$ if and only if $a = b$ is provable in $\mathcal{L}$. In this way, one can prove the following *CCC completeness theorem for λ-calculus*.

Proposition 6.17. *For any theory $\mathcal{L}$ in the λ-calculus, one has the following:*

1. *For any terms a, b, $\mathcal{L} \vdash a = b$ iff for all models M in CCCs, $[\![a]\!]_M = [\![b]\!]_M$.*
2. *Moreover, for any type A, there is a closed $t : A$ iff for all models M in CCCs, there is an arrow $1 \to [\![A]\!]_M$.*

This proposition says that the λ-calculus is *deductively sound and complete* for models in CCCs. It is worth emphasizing that completeness is not true if one restricts attention to models in the single category **Sets**; indeed, there are many examples of theories in λ-calculus in which equations holding for all models in **Sets** are still not provable (see the exercises for an example).

Soundness (i.e., the "only if" direction of the above statements) follows from the following UMP of the cartesian closed category $\mathbf{C}(\mathcal{L})$, analogous to the one for any algebra presented by generators and relations. Given any model M of $\mathcal{L}$ in any cartesian closed category $\mathbf{C}$, there is a unique functor,

$$[\![-]\!]_M : \mathbf{C}(\mathcal{L}) \to \mathbf{C}$$

preserving the CCC structure, given by

$$[\![X]\!]_M = M$$

for the basic type X, and similarly for the other basic types and terms of $\mathcal{L}$. In this precise sense, the theory $\mathcal{L}$ is a presentation of the cartesian closed category $\mathbf{C}(\mathcal{L})$ by generators and relations.

Finally, let us note that the notions of λ-calculus and CCC are essentially "equivalent," in the sense that any cartesian closed category $\mathbf{C}$ also gives rise to a λ-calculus $\mathcal{L}(\mathbf{C})$, and this construction is essentially inverse to the one just sketched.

Briefly, given $\mathbf{C}$, we define $\mathcal{L}(\mathbf{C})$ by

- Basic types: the objects of $\mathbf{C}$
- Basic terms: $a : A \to B$ for each $a : A \to B$ in $\mathbf{C}$
- Equations: many equations identifying the λ-calculus operations with the corresponding category and CCC structure on $\mathbf{C}$, for example,

$$\lambda x.\mathrm{fst}(x) = p_1$$
$$\lambda x.\mathrm{snd}(x) = p_2$$
$$\lambda y.f(x, y) = \tilde{f}(x)$$
$$g(f(x)) = (g \circ f)(x)$$
$$\lambda y.y = 1_A$$

This suffices to ensure that there is an isomorphism of categories,

$$\mathbf{C}(\mathcal{L}(\mathbf{C})) \cong \mathbf{C}.$$

Moreover, the theories $\mathcal{L}$ and $\mathcal{L}(\mathbf{C}(\mathcal{L}))$ are also "equivalent" in a suitable sense, involving the kinds of considerations typical of comparing different presentations of algebras. We refer the reader to the excellent book by Lambek and Scott (1986), for further details.

6.7　Variable sets

We conclude with a special kind of CCC related to the so-called Kripke models of logic, namely categories of *variable sets*. These categories provide specific examples of the "algebraic" semantics of IPC and λ-calculus just given.

6.7.1　IPC

Let us begin by very briefly reviewing the notion of a Kripke model of IPC from our algebraic point of view; we focus on the positive fragment involving only $\top, p \wedge q, p \Rightarrow q$, and variables.

A *Kripke model* of this language $\mathcal{L}$ consists of a poset I of "possible worlds," which we write $i \leq j$, together with a relation between worlds i and propositions p,

$$i \Vdash p,$$

read "p holds at i." This relation is assumed to satisfy the following conditions:

(1) $i \Vdash p$ and $i \leq j$ implies $j \Vdash p$

(2) $i \Vdash \top$

(3) $i \Vdash p \wedge q$ iff $i \Vdash p$ and $i \Vdash q$

(4) $i \Vdash p \Rightarrow q$ iff $j \Vdash p$ implies $j \Vdash q$ for all $j \geq i$.

One then sets

$$I \Vdash p \quad \text{iff} \quad i \Vdash p \quad \text{for all } i \in I.$$

And finally, we have the well-known theorem,

Theorem 6.18 (Kripke completeness for IPC). *A propositional formula p is provable from the rules for* IPC *iff it holds in all Kripke models, that is, iff $I \Vdash p$ for all relations $\Vdash$ over all posets I,*

$$\text{IPC} \vdash p \quad \textit{iff} \quad I \Vdash p \quad \textit{for all } I.$$

Now let us see how to relate this result to our formulation of the semantics of IPC in Heyting algebras. First, the relation $\Vdash \subseteq I \times \mathrm{Prop}(\mathcal{L})$ between worlds I and propositional formulas $\mathrm{Prop}(\mathcal{L})$ can be equivalently formulated as a mapping,

$$[\![-]\!] : \mathrm{Prop}(\mathcal{L}) \longrightarrow \mathbf{2}^I, \tag{6.7}$$

where we write $\mathbf{2}^I = \mathrm{Hom}_{\mathbf{Pos}}(I, \mathbf{2})$ for the exponential poset of monotone maps from I into the poset $\mathbf{2} = \{\bot \leq \top\}$. This poset is a CCC, and indeed a Heyting algebra, the proof of which we leave as an exercise for the reader. The mapping (6.7) is determined by the condition

$$[\![p]\!](i) = \top \quad \text{iff} \quad i \Vdash p.$$

Now, in terms of the Heyting algebra semantics of IPC developed in Section 6.4 (adapted in the evident way to the current setting without the coproducts $\bot, p \vee q$, and writing HA^- for Heyting algebras without coproducts, i.e., poset CCCs), the poset $\mathrm{HA}^-(\mathcal{L})$ is a quotient of $\mathrm{Prop}(\mathcal{L})$ by the equivalence relation of mutual derivability $p \dashv\vdash q$, which clearly makes it a CCC, and the map (6.7) therefore determines a model (with the same name),

$$[\![-]\!] : \mathrm{HA}^-(\mathcal{L}) \longrightarrow \mathbf{2}^I.$$

Indeed, condition (1) above ensures that $[\![p]\!] : I \to \mathbf{2}$ is monotone, and (2)–(4) ensure that $[\![-]\!]$ is a homomorphism of poset CCCs, that is, that it is monotone and preserves the CCC structure (exercise!). Thus, a Kripke model is just an "algebraic" model in a Heyting algebra of the special form $\mathbf{2}^I$. The Kripke completeness theorem for positive IPC above then follows from Heyting-valued completeness theorem proposition 6.14 together with the following, purely algebraic, embedding theorem for poset CCCs.

Proposition 6.19. *For every poset CCC* $\mathbf{A}$*, there is a poset* I *and an injective, monotone map,*

$$y : \mathbf{A} \rightarrowtail \mathbf{2}^I,$$

preserving CCC structure.

Proof. We can take $I = \mathbf{A}^{\mathrm{op}}$ and $y(a) : \mathbf{A}^{\mathrm{op}} \to \mathbf{2}$, the "truth-value" of $x \leq a$, that is, $y(a)$ is determined by

$$y(a)(x) = \top \quad \text{iff} \quad x \leq a.$$

Clearly, $y(a)$ is monotone and contravariant, while y itself is monotone and covariant. We leave it as an exercise to verify that y is injective and preserves the CCC structure, but note that $\mathbf{2}^{\mathbf{A}^{\mathrm{op}}}$ can be identified with the collection of *lower sets* $S \subseteq \mathbf{A}$ in $\mathbf{A}$, that is, subsets that are closed downward: $x \leq y \in S$ implies $x \in S$. Under this identification, we then have $y(a) = {\downarrow}(a) = \{x \mid x \leq a\}$.

A proof is also given in Chapter 8 as a consequence of the Yoneda lemma. $\square$

The result can be extended from poset CCCs to Heyting algebras, thus recovering the usual Kripke completeness theorem for full IPC, by the same argument using a more delicate embedding theorem that also preserves the coproducts $\bot$ and $p \vee q$.

6.7.2 λ-calculus

We now want to generalize the foregoing from propositional logic to the λ-calculus, motivated by the insight that the latter is the proof theory of the former (according to the Curry–Howard–correspondence). Categorically spea-king, we are generalizing from the poset case to the general case of a CCC. According to the "propositions-as-types" conception behind the C–H correspon-dence, we therefore should replace the poset CCC of idealized propositions **2** with the general CCC of idealized types **Sets**. We therefore model the λ-calculus in categories of the form **Sets**^{I} for posets I, which can be regarded as comprised of "I-indexed," or "variable sets," as we now indicate.

Given a poset I, an *I-indexed set* is a family of sets $(A_i)_{i \in I}$ together with transition functions $\alpha_{ij} : A_i \to A_j$ for each $i \leq j$, satisfying the compatibility conditions:

- $\alpha_{ik} = \alpha_{jk} \circ \alpha_{ij}$ whenever $i \leq j \leq k$,
- $\alpha_{ii} = 1_{A_i}$ for all i.

In other words, it is simply a functor,

$$A : I \longrightarrow \textbf{Sets}.$$

We can think of such I-indexed sets as "sets varying in a parameter" from the poset I. For instance, if $I = \mathbb{R}$ thought of as time, then an $\mathbb{R}$-indexed set A may be thought of as a set varying through time: some elements $a, b \in A_t$ may become identified over time (the αs need not be injective), and new elements may appear over time (the αs need not be surjective), but once an element is in the set ($a \in A_t$), it stays in forever ($\alpha_{tt'}(a) \in A_{t'}$). For a more general poset I, the variation is parameterized accordingly.

A product of two variable sets A and B can be constructed by taking the pointwise products $(A \times B)(i) = A(i) \times B(i)$ with the evident transition maps,

$$\alpha_{ij} \times \beta_{ij} : A(i) \times B(i) \to A(j) \times B(j) \qquad i \leq j$$

where $\beta_{ij} : B_i \to B_j$ is the transition map for B. This plainly gives an I-indexed set, but to check that it really is a product we need to make **Sets**^{I} into a category and verify the UMP (respectively, the operations and equations of Section 6.5). What is a map of I-indexed sets $f : A \to B$? One natural proposal is this: it is an I-indexed family of functions $(f_i : A_i \to B_i)_{i \in I}$ that are compatible with the transition maps, in the sense that whenever $i \leq j$, then the following commutes:

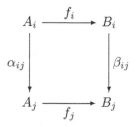

We can think of this condition as saying that f takes elements $a \in A$ to elements $f(a) \in B$ without regard to when the transition is made, since given $a \in A_i$ it does not matter if we first wait until $j \geq i$ and then take $f_j(\alpha_{ij}(a))$, or go right away to $f_i(a)$ and then wait until $\beta_{ij}(f_i(a))$.

Indeed, in Chapter 7 we see that this type of map is exactly what is called a "natural transformation" of the functors A and B. These maps $f : A \to B$ compose in the evident way:

$$(g \circ f)_i = g_i \circ f_i : A_i \longrightarrow B_i$$

to make $Sets^I$ into a category, the *category of I-indexed sets*. It is now an easy exercise to confirm that the specification of the product $A \times B$ just given really is a product in the resulting category $\mathbf{Sets}^I$, and the terminal object is obviously the constant index set 1, so $\mathbf{Sets}^I$ has all finite products.

What about exponentials? The first attempt at defining pointwise exponentials,

$$(B^A)_i = B_i^{A_i}$$

fails, because the indexing is covariant in B and *contra*variant in A, as the reader should confirm. The idea that maybe B^A is just the collection of all index maps from A to B also fails, because it is not indexed! The solution is a combintion of these two ideas which generalizes the "Kripke" exponential as follows. For each $i \in I$, let

$$\uparrow(i) \subseteq I$$

be the upper set above i, regarded as a subposet. Then for any $A : I \to \mathbf{Sets}$, let $A|_i$ be the restriction,

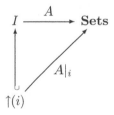

This determines an indexed set over $\uparrow(i)$. Given any $f : A \to B$ and $i \in I$, there is an evident restriction $f|_i : A|_i \to B|_i$ which is defined to be simply $(f|_i)_j = f_j$ for any $j \geq i$. Now we can define

$$(B^A)_i = \{f : A|_i \to B|_i \mid f \text{ is } \uparrow(i)\text{-indexed}\}$$

with the transition maps given by

$$f \mapsto f|_j \qquad j \geq i$$

It is immediate that this determines an I-indexed set B^A. That it is actually the exponential of A and B in $\mathbf{Sets}^I$ is shown later, as an easy consequence of the Yoneda lemma. For the record, we therefore have the following (proof deferred).

Proposition 6.20. *For any poset I, the category $Sets^I$ of I-indexed sets and functions is cartesian closed.*

Definition 6.21. A *Kripke model* of a theory $\mathcal{L}$ in the λ-calculus is a model (in the sense of definition 6.16) in a cartesian closed category of the form $\mathbf{Sets}^I$ for a poset I.

For instance, it can be seen that a Kripke model over a poset I of a conventional algebraic theory such as the theory of groups is just an I-indexed group, that is, a functor $I \to \mathbf{Group}$. In particular, if $I = \mathcal{O}(X)^{\mathrm{op}}$ for a topological space X, then this is just what the topologist calls a "presheaf of groups." On the other hand, it also agrees with (or generalizes) the logician's notion of a Kripke model of a first-order language, in that it consists of a varying domain of "individuals" equipped with varying structure.

Finally, in order to generalize the Kripke completeness theorem for IPC to λ-calculus, it clearly suffices to sharpen our general CCC completeness theorem, proposition 6.17, to the special models in CCCs of the form $\mathbf{Sets}^I$ by means of an embedding theorem analogous to proposition 6.19. Indeed, one can prove the following.

Proposition 6.22. *For every CCC* $\mathbf{C}$*, there is a poset I and a functor,*

$$y : \mathbf{C} \rightarrowtail \mathbf{Sets}^I,$$

that is injective on both objects and arrows and preserves CCC structure. Moreover, every map between objects in the image of y is itself in the image of y (y is "full").

The full proof of this result involves methods from topos theory that are beyond the scope of this book. But a significant part of it, to be given below, is entirely analogous to the proof of the poset case, and will again be a consequence of the Yoneda lemma.

6.8   Exercises

1. Show that for all finite sets M and N,

$$|N^M| = |N|^{|M|},$$

   where $|K|$ is the number of elements in the set K, while N^M is the exponential in the category of sets (the set of all functions $f : M \to N$), and n^m is the usual exponentiation operation of arithmetic.

2. Show that for any three objects A, B, C in a cartesian closed category, there are isomorphisms:

   (a) $(A \times B)^C \cong A^C \times B^C$

   (b) $(A^B)^C \cong A^{B \times C}$

3. Determine the exponential transpose $\tilde{\varepsilon}$ of evaluation $\varepsilon : B^A \times A \to B$ (for any objects in any CCC). In **Sets**, determine the transpose $\tilde{1}$ of the identity

$1 : A \times B \to A \times B$. Also determine the transpose of $\varepsilon \circ \tau : A \times B^A \to B$, where $\tau : A \times B^A \to B^A \times A$ is the "twist" arrow $\tau = \langle p_2, p_1 \rangle$.

4. Is the category of monoids cartesian closed?

5. Verify the description given in the text of the exponential graph H^G for two graphs G and H. Determine the exponential $\mathbf{2}^G$, where $\mathbf{2}$ is the graph $v_1 \to v_2$ with two vertices and one edge, and G is an arbitrary graph. Determine $\mathbf{2}^G$ explicitly for G the graph pictured below.

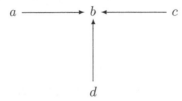

6. Consider the category of sets equipped with a (binary) relation, $(A, R \subseteq A \times A)$, with maps $f : (A, R) \to (B, S)$ being those functions $f : A \to B$ such that aRa' implies $f(a)Sf(a')$. Show this category is cartesian closed by describing it as a subcategory of graphs.

7. Consider the category of sets equipped with a distinguished subset, $(A, P \subseteq A)$, with maps $f : (A, P) \to (B, Q)$ being those functions $f : A \to B$ such that $a \in P$ iff $f(a) \in Q$. Show this category is cartesian closed by describing it as a category of pairs of sets.

8. Consider the category of "pointed sets," that is, sets equipped with a distinguished element, $(A, a \in A)$, with maps $f : (A, a) \to (B, b)$ being those functions $f : A \to B$ such that $f(a) = b$. Is this category cartesian closed?

9. Show that for any objects A, B in a cartesian closed category, there is a bijective correspondence between points of the exponential $1 \to B^A$ and arrows $A \to B$.

10. Show that the category of ωCPOs is cartesian closed, but that the category of *strict* ωCPOs is not (the strict ωCPOs are the ones with initial object $\bot$, and the continuous maps between them are supposed to preserve $\bot$).

11. (a) Show that in any cartesian closed poset with joins $p \vee q$, the following "distributive" law of IPC holds:

$$((p \vee q) \Rightarrow r) \Rightarrow ((p \Rightarrow r) \wedge (q \Rightarrow r))$$

(b) Generalize the foregoing problem to an arbitrary category (not necessarily a poset), by showing that there is always an arrow of the corresponding form.

(c) If you are brave, show that the previous two arrows are isomorphisms.

12. Prove that in a CCC $\mathbf{C}$, exponentiation with a fixed base object C is a contravariant functor $C^{(-)} : \mathbf{C}^{\mathrm{op}} \to \mathbf{C}$, where $C^{(-)}(A) = C^A$.

13. Show that in a cartesian closed category with coproducts, the products necessarily distribute over the coproducts,

$$(A \times C) + (B \times C) \cong (A + B) \times C.$$

14. In the λ-calculus, consider the theory (due to Dana Scott) of a *reflexive domain*: there is one basic type D, two constants s and r of types

$$s : (D \to D) \to D$$
$$r : D \to (D \to D),$$

and two equations,

$$srx = x \quad (x : D)$$
$$rsy = y \quad (y : D \to D).$$

Prove that, up to isomorphism, this theory has only one model M in **Sets**, and that *every* equation holds in M.

15. Complete the proof from the text of Kripke completeness for the positive fragement of IPC as follows:

   (a) Show that for any poset I, the exponential poset $\mathbf{2}^I$ is a Heyting algebra. (Hint: the limits and colimits are "pointwise," and the Heyting implication $p \Rightarrow q$ is defined at $i \in I$ by $(p \Rightarrow q)(i) = \top$ iff for all $j \geq i, p(j) \leq q(j)$.)

   (b) Show that for any poset CCC **A**, the map $y : \mathbf{A} \to \mathbf{2}^{\mathbf{A}^{\mathrm{op}}}$ defined in the text is indeed (i) monotone, (ii) injective, and (iii) preserves CCC structure.

16. Verify the claim in the text that the products $A \times B$ in categories **Sets**^{I} of I-indexed sets (I a poset) can be computed "pointwise." Show, moreover, that the same is true for all limits and colimits.

7

NATURALITY

We now want to start considering categories and functors more systematically, developing the "category theory" of category theory itself, rather than of other mathematical objects, like groups, or formulas in a logical system. Let me emphasize that, while some of this may look a bit like "abstract nonsense," the idea behind it is that when one has a particular application at hand, the theory can then be specialized to that concrete case. The notion of a functor is a case in point; developing its general theory makes it a clarifying, simplifying, and powerful tool in its many instances.

7.1 Category of categories

We begin by reviewing what we know about the category **Cat** of categories and functors and tying up some loose ends.

We have already seen that **Cat** has finite coproducts $\mathbf{0}$, $\mathbf{C} + \mathbf{D}$; and finite products $\mathbf{1}$, $\mathbf{C} \times \mathbf{D}$. It is very easy to see that there are also all small coproducts and products, constructed analogously. We can therefore show that **Cat** has all limits by constructing equalizers. Thus, let categories **C** and **D** and parallel functors F and G be given, and define the category **E** and functor E,

$$\mathbf{E} \xrightarrow{\ E\ } \mathbf{C} \underset{G}{\overset{F}{\rightrightarrows}} \mathbf{D}$$

as follows (recall that for a category **C**, we write $\mathbf{C}_0$ and $\mathbf{C}_1$ for the collections of objects and arrows, respectively):

$$\mathbf{E}_0 = \{C \in \mathbf{C}_0 \mid F(C) = G(C)\}$$
$$\mathbf{E}_1 = \{f \in \mathbf{C}_1 \mid F(f) = G(f)\}$$

and let $E : \mathbf{E} \to \mathbf{C}$ be the evident inclusion. This is then an equalizer, as the reader can easily check.

The category **E** is an example of a *subcategory*, that is, a monomorphism in **Cat** (recall that equalizers are monic). Often, by a subcategory of a category **C** one means specifically a collection **U** of some of the objects and arrows, $\mathbf{U}_0 \subseteq \mathbf{C}_0$ and $\mathbf{U}_1 \subseteq \mathbf{C}_1$), that is closed under the operations dom, cod, id, and $\circ$. There is then an evident inclusion functor

$$i : \mathbf{U} \to \mathbf{C}$$

which is clearly monic.

In general, coequalizers of categories are more complicated to describe—indeed, even for posets, determining the coequalizer of a pair of monotone maps can be quite involved, as the reader should consider.

There are various properties of functors other than being monic and epic that turn out to be quite useful in **Cat**. A few of these are given by the following:

Definition 7.1. A functor $F : \mathbf{C} \to \mathbf{D}$ is said to be

- *injective on objects* if the object part $F_0 : \mathbf{C}_0 \to \mathbf{D}_0$ is injective, it is *surjective on objects* if F_0 is surjective.

- Similarly, F is *injective* (resp. *surjective*) *on arrows* if the arrow part $F_1 : \mathbf{C}_1 \to \mathbf{D}_1$ is injective (resp. surjective).

- F is *faithful* if for all $A, B \in \mathbf{C}_0$, the map

$$F_{A,B} : \mathrm{Hom}_{\mathbf{C}}(A, B) \to \mathrm{Hom}_{\mathbf{D}}(FA, FB)$$

  defined by $f \mapsto F(f)$ is injective.

- Similarly, F is *full* if $F_{A,B}$ is always surjective.

What is the difference between being faithful and being injective on arrows? Consider, for example, the "codiagonal functor" $\nabla : \mathbf{C} + \mathbf{C} \to \mathbf{C}$, as indicated in the following:

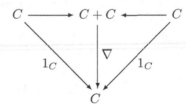

∇ is faithful, but not injective on arrows.

A *full subcategory*

$$\mathbf{U} \rightarrowtail \mathbf{C}$$

consists of some objects of $\mathbf{C}$ and *all* of the arrows between them (thus satisfying the closure conditions for a subcategory). For example, the inclusion functor $\mathbf{Sets}_{fin} \rightarrowtail \mathbf{Sets}$ is full and faithful, but the forgetful functor $\mathbf{Groups} \rightarrowtail \mathbf{Sets}$ is faithful but not full.

Example 7.2. There is another "forgetful" functor for groups, namely to the category **Cat** of categories,

$$G : \mathbf{Groups} \to \mathbf{Cat}.$$

Observe that this functor is full and faithful, since a functor between groups $F : G(A) \to G(B)$ is exactly the same thing as a group homomorphism.

And exactly the same situation holds for monoids.

For posets, too, there is a full and faithful, forgetful functor

$$P : \textbf{Pos} \to \textbf{Cat}$$

again because a functor between posets $F : P(A) \to P(B)$ is exactly a monotone map. And the same thing holds for the "discrete category" functor $S : \textbf{Sets} \to \textbf{Cat}$.

Thus, **Cat** provides a setting for *comparing structures of many different kinds*. For instance, one can have a functor $R : G \to \textbf{C}$ from a group G to a category **C** that is not a group. If **C** is a poset, then any such functor must be trivial (why?). But if **C** is, say, the category of finite dimensional, real vector spaces and linear maps, then a functor R is exactly a *linear representation* of the group G, representing every element of G as an invertible matrix of real numbers and the group multiplication as matrix multiplication.

What is a functor $g : P \to G$ from a poset to a group? Since G has only one object $*$, it has $g(p) = * = g(q)$ for all $p, q \in P$. For each $p \leq q$, it picks an element $g_{p,q}$ in such a way that

$$g_{p,p} = u \qquad \text{(the unit of } G\text{)}$$

$$g_{q,r} \cdot g_{p,q} = g_{p,r}.$$

For example, take $P = (\mathbb{R}, \leq)$ to be the ordered real numbers and $G = (\mathbb{R}, +)$ the additive group of reals, then *subtraction* is a functor,

$$g : (\mathbb{R}, \leq) \to (\mathbb{R}, +)$$

defined by

$$g_{x,y} = (y - x).$$

Indeed, we have

$$g_{x,x} = (x - x) = 0$$

$$g_{y,z} \cdot g_{x,y} = (z - y) + (y - x) = (z - x) = g_{x,z}.$$

7.2   Representable structure

Let **C** be a *locally small category*, so that we have the *representable functors*,

$$\text{Hom}_{\textbf{C}}(C, -) : \textbf{C} \to \textbf{Sets}$$

for all objects $C \in \textbf{C}$. This functor is evidently faithful if the object C has the property that for any objects X and Y and arrows $f, g : X \rightrightarrows Y$, if $f \neq g$ there is an arrow $x : C \to X$ such that $fx \neq gx$. That is, the arrows in the category are distinguished by their effect on generalized elements based at C. Such an object C is called a *generator* for **C**.

In the category of sets, for example, the terminal object 1 is a generator. In groups, as we have already discussed, the free group $F(1)$ on one element is a generator. Indeed, the functor represented by $F(1)$ is isomorphic to the forgetful functor $U : \mathbf{Grp} \to \mathbf{Sets}$,

$$\mathrm{Hom}(F(1), G) \cong U(G). \tag{7.1}$$

This isomorphism not only holds for each group G, but also respects group homomorphisms, in the sense that for any such $h : G \to H$, there is a commutative square,

$$
\begin{array}{ccc}
G & \mathrm{Hom}(F(1), G) \xrightarrow{\;\cong\;} U(G) \\[4pt]
{\scriptstyle h}\downarrow & \quad {\scriptstyle h_*}\downarrow \qquad\qquad \downarrow{\scriptstyle U(h)} \\[4pt]
H & \mathrm{Hom}(F(1), H) \xrightarrow[\cong]{} U(H)
\end{array}
$$

One says that the isomorphism (7.1) is "natural in G." In a certain sense, this also "explains" why the forgetful functor U preserves all limits, since representable functors necessarily do. The related fact that the forgetful functor is faithful is a precise way to capture the vague idea, which we initially used for motivation, that the category of groups is "concrete."

Recall that there are also *contravariant* representable functors

$$\mathrm{Hom}_{\mathbf{C}}(-, C) : \mathbf{C}^{\mathrm{op}} \to \mathbf{Sets}$$

taking $f : A \to B$ to $f^* : \mathrm{Hom}_{\mathbf{C}}(B, C) \to \mathrm{Hom}_{\mathbf{C}}(A, C)$ by $f^*(h) = h \circ f$ for $h : B \to C$.

Example 7.3. Given a group G in a (locally small) category **C**, the contravariant representable functor $\mathrm{Hom}_{\mathbf{C}}(-, G)$ actually has a group structure, giving a functor

$$\mathrm{Hom}_{\mathbf{C}}(-, G) : \mathbf{C}^{\mathrm{op}} \to \mathbf{Grp}.$$

In **Sets**, for example, for each set X, we can define the operations on the group $\mathrm{Hom}(X, G)$ pointwise,

$$u(x) = u \quad \text{(the unit of } G)$$
$$(f \cdot g)(x) = f(x) \cdot g(x)$$
$$f^{-1}(x) = f(x)^{-1}.$$

In this case, we have an isomorphism

$$\mathrm{Hom}(X, G) \cong \Pi_{x \in X} G$$

with the product group. Functoriality in X is given simply by precomposition; thus, for any function $h : Y \rightarrow X$, one has

$$
\begin{aligned}
h^*(f \cdot g)(y) &= (f \cdot g)(h(y)) \\
&= f(h(y)) \cdot g(h(y)) \\
&= h^*(f)(y) \cdot h^*(g)(y) \\
&= (h^*(f) \cdot h^*(g))(y)
\end{aligned}
$$

and similarly for inverses and the unit. Indeed, it is easy to see that this construction works just as well for any other algebraic structure defined by operations and equations. Nor is there anything special about the category **Sets** here; we can do the same thing in any category with an internal algebraic structure.

For instance, in topological spaces, one has the ring $\mathbb{R}$ of real numbers and, for any space X, the ring

$$
\mathcal{C}(X) = \mathrm{Hom}_{\mathrm{Top}}(X, \mathbb{R})
$$

of real-valued, continuous functions on X. Just as in the previous case, if

$$
h : Y \rightarrow X
$$

is any continuous function, we then get a ring homomorphism

$$
h^* : \mathcal{C}(X) \rightarrow \mathcal{C}(Y)
$$

by precomposing with h. The recognition of $\mathcal{C}(X)$ as representable ensures that this "ring of real-valued functions" construction is functorial,

$$
\mathcal{C} : \mathbf{Top}^{\mathrm{op}} \rightarrow \mathbf{Rings}.
$$

Note that in passing from $\mathbb{R}$ to $\mathrm{Hom}_{\mathbf{Top}}(X, \mathbb{R})$, all the algebraic structure of $\mathbb{R}$ is retained, but properties determined by conditions that are not strictly equational are not necessarily preserved. For instance, $\mathbb{R}$ is not only a ring, but also a *field*, meaning that every nonzero real number r has a multiplicative inverse r^{-1}; formally,

$$
\forall x (x = 0 \vee \exists y.\ y \cdot x = 1).
$$

To see that this condition fails in, for example, $\mathcal{C}(\mathbb{R})$, consider the continuous function $f(x) = x^2$. For any argument $y \neq 0$, the multiplicative inverse must be $g(y) = 1/y^2$. But if this function were to be continuous, at 0 it would have to be $\lim_{y \to 0} 1/y^2$ which does not exist in $\mathbb{R}$.

Example 7.4. A very similar situation occurs in the category **BA** of Boolean algebras. Given the Boolean algebra **2** with the usual (truth-table) operations $\wedge, \vee, \neg, 0, 1$, for any set X, we make the set

$$
\mathrm{Hom}_{\mathbf{Sets}}(X, \mathbf{2})
$$

into a Boolean algebra with the pointwise operations:

$$0(x) = 0$$
$$1(x) = 1$$
$$(f \wedge g)(x) = f(x) \wedge g(x)$$

etc.

When we define the operations in this way in terms of those on $\mathbf{2}$, we see immediately that $\text{Hom}(X, \mathbf{2})$ is a Boolean algebra too, and that precomposition is a contravariant functor,

$$\text{Hom}(-, \mathbf{2}) : \mathbf{Sets}^{\text{op}} \to \mathbf{BA}$$

into the category $\mathbf{BA}$ of Boolean algebras and their homomorphisms.

Now observe that for any set X, the familiar isomorphism

$$\text{Hom}(X, \mathbf{2}) \cong \mathcal{P}(X)$$

between characteristic functions $\phi : X \to \mathbf{2}$ and subsets $V_\phi = \phi^{-1}(1) \subseteq X$, relates the pointwise Boolean operations in $\text{Hom}(X, \mathbf{2})$ to the subset operations of intersection, union, etc. in $\mathcal{P}(X)$:

$$V_{\phi \wedge \psi} = V_\phi \cap V_\psi$$
$$V_{\phi \vee \psi} = V_\phi \cup V_\psi$$
$$V_{\neg \phi} = X - V_\phi$$
$$V_1 = X$$
$$V_0 = \emptyset$$

In this sense, the set-theoretic Boolean operations on $\mathcal{P}(X)$ are induced by those on $\mathbf{2}$, and the powerset $\mathcal{P}$ is seen to be a contravariant functor to the category of Boolean algebras,

$$\mathcal{P}^{\text{BA}} : \mathbf{Sets}^{\text{op}} \to \text{BA}.$$

As was the case for the covariant representable functor $\text{Hom}_{\mathbf{Grp}}(F(1), -)$ and the forgetful functor U from groups to sets, here the contravariant functors $\text{Hom}_{\mathbf{Sets}}(-, \mathbf{2})$ and $\mathcal{P}^{\text{BA}}$ from sets to Boolean algebras can also be seen to be *naturally* isomorphic, in the sense that for any function $f : Y \to X$, the following square of Boolean algebras and homomorphisms commutes:

$$
\begin{array}{ccc}
X & \text{Hom}(X, \mathbf{2}) \xrightarrow{\ \cong\ } \mathcal{P}(X) \\
\Big\uparrow f & \Big\downarrow f^* \qquad\qquad \Big\downarrow f^{-1} \\
Y & \text{Hom}(Y, \mathbf{2}) \xrightarrow[\cong]{} \mathcal{P}(Y)
\end{array}
$$

7.3 Stone duality

Before considering the topic of naturality more systematically, let us take a closer look at the foregoing example of powersets and Boolean algebras.

Recall that an *ultrafilter* in a Boolean algebra B is a proper subset $U \subset B$ such that

- $1 \in U$
- $x, y \in U$ implies $x \wedge y \in U$
- $x \in U$ and $x \leq y$ implies $y \in U$
- if $U \subset U'$ and U' is a filter, then $U' = B$

The maximality condition on U is equivalent to the condition that for every $x \in B$, either $x \in U$ or $\neg x \in U$ but not both (exercise!).

We already know that there is an isomorphism between the set $\mathrm{Ult}(B)$ of ultrafilters on B and the Boolean homomorphisms $B \to \mathbf{2}$,

$$\mathrm{Ult}(B) \cong \mathrm{Hom}_{\mathbf{BA}}(B, \mathbf{2}).$$

This assignment $\mathrm{Ult}(B)$ is functorial and contravariant, and the displayed isomorphism above is natural in B. Indeed, given a Boolean homomorphism $h : B' \to B$, let

$$\mathrm{Ult}(h) = h^{-1} : \mathrm{Ult}(B) \to \mathrm{Ult}(B').$$

Of course, we have to show that the inverse image $h^{-1}(U) \subseteq B'$ of an ultrafilter $U \subset B$ is an ultrafilter in B'. But since we know that $U = \chi_U^{-1}(1)$ for some $\chi_U : B \to 2$, we have

$$\mathrm{Ult}(h)(U) = h^{-1}(\chi_U^{-1}(1))$$
$$= (\chi_U \circ h)^{-1}(1).$$

Therefore, $\mathrm{Ult}(h)(U)$ is also an ultrafilter. Thus, we have a contravariant functor of ultrafilters

$$\mathrm{Ult} : \mathbf{BA}^{\mathrm{op}} \to \mathbf{Sets},$$

as well as the contravariant powerset functor coming back

$$\mathcal{P}^{\mathbf{BA}} : \mathbf{Sets}^{\mathrm{op}} \to \mathbf{BA}.$$

The constructions,

$$\mathbf{BA}^{\mathrm{op}} \xleftarrow[\mathrm{Ult}]{(\mathcal{P}^{\mathbf{BA}})^{\mathrm{op}}} \mathbf{Sets}$$

are *not* mutually inverse, however. For in general, $\mathrm{Ult}(\mathcal{P}(X))$ is much larger than X, since there are many ultrafilters in $\mathcal{P}(X)$ that are not "principal," that is, of the form $\{U \subseteq X \mid x \in U\}$ for some $x \in X$. (But what if X is finite?) Instead, there is a more subtle relation between these functors that we consider in more detail later; namely, these are an example of adjoint functors.

For now, consider the following observations. Let

$$\mathcal{U} = \mathrm{Ult} \circ (\mathcal{P}^{\mathbf{BA}})^{\mathrm{op}} : \mathbf{Sets} \to \mathbf{BA}^{\mathrm{op}} \to \mathbf{Sets}$$

so that

$$\mathcal{U}(X) = \{U \subseteq \mathcal{P}(X) \mid U \text{ is an ultrafilter}\}$$

is a *covariant* functor on **Sets**. Now, observe that for any set X, there is a function

$$\eta : X \to \mathcal{U}(X)$$

taking each element $x \in X$ to the principal ultrafilter

$$\eta(x) = \{U \subseteq X \mid x \in U\}.$$

This map is "natural" in X, that is, for any function $f : X \to Y$, the following diagram commutes:

$$
\begin{array}{ccc}
X & \xrightarrow{\;\eta_X\;} & \mathcal{U}(X) \\
{\scriptstyle f}\downarrow & & \downarrow{\scriptstyle \mathcal{U}(f)} \\
Y & \xrightarrow[\;\eta_Y\;]{} & \mathcal{U}(Y)
\end{array}
$$

This is so because, for any ultrafilter $\mathcal{V}$ in $\mathcal{P}(X)$,

$$\mathcal{U}(f)(\mathcal{V}) = \{U \subseteq Y \mid f^{-1}(U) \in \mathcal{V}\}.$$

So in the case of the principal ultrafilters $\eta(x)$, we have

$$
\begin{aligned}
(\mathcal{U}(f) \circ \eta_X)(x) &= \mathcal{U}(f)(\eta_X(x)) \\
&= \{V \subseteq Y \mid f^{-1}(V) \in \eta_X(x)\} \\
&= \{V \subseteq Y \mid x \in f^{-1}(V)\} \\
&= \{V \subseteq Y \mid fx \in V\} \\
&= \eta_Y(fx) \\
&= (\eta_Y \circ f)(x).
\end{aligned}
$$

Finally, observe that there is an analogous natural map at the "other side" of this situation, in the category of Boolean algebras. Specifically, for every Boolean algebra B, there is a homomorphism similar to the function η,

$$\phi_B : B \to \mathcal{P}(\mathrm{Ult}(B))$$

given by

$$\phi_B(b) = \{\mathcal{V} \in \text{Ult}(B) \mid b \in \mathcal{V}\}.$$

It is not hard to see that ϕ_B is always injective. For, given any distinct elements $b, b' \in B$, the Boolean prime ideal theorem implies that there is an ultrafilter $\mathcal{V}$ containing one but not the other. The Boolean algebra $\mathcal{P}(\text{Ult}(B))$, together with the homomorphism ϕ_B, is called the *Stone representation* of B. It presents the arbitrary Boolean algebra B as an algebra of subsets. For the record, we thus have the following step toward a special case of the far-reaching *Stone duality theorem*.

Proposition 7.5. *Every Boolean algebra B is isomorphic to one consisting of subsets of some set X, equipped with the set-theoretical Boolean operations.*

7.4 Naturality

A natural transformation is a morphism of functors. That is right: for fixed categories $\mathbf{C}$ and $\mathbf{D}$, we can regard the functors $\mathbf{C} \to \mathbf{D}$ as the *objects* of a new category, and the arrows between these objects are what we are going to call natural transformations. They are to be thought of as different ways of "relating" functors to each other, in a sense that we now explain.

Let us begin by considering a certain kind of situation that often arises: we have some "construction" on a category $\mathbf{C}$ and some other "construction," and we observe that these two "constructions" are related to each other in a way that is independent of the specific objects and arrows involved. That is, the relation is really between the constructions themselves. To give a simple example, suppose $\mathbf{C}$ has products and consider, for objects $A, B, C \in \mathbf{C}$,

$$(A \times B) \times C \quad \text{and} \quad A \times (B \times C).$$

Regardless of what objects A, B, and C are, we have an isomorphism

$$h : (A \times B) \times C \xrightarrow{\sim} A \times (B \times C).$$

What does it mean that this isomorphism does not really depend on the particular objects A, B, C? One way to explain it is this:

Given any $f : A \to A'$, we get a commutative square

$$
\begin{array}{ccc}
(A \times B) \times C & \xrightarrow{\ h_A\ } & A \times (B \times C) \\
\Big\downarrow & & \Big\downarrow \\
(A' \times B) \times C & \xrightarrow[h_{A'}]{} & A' \times (B \times C)
\end{array}
$$

So what we really have is an isomorphism between the "constructions"

$$(- \times B) \times C \quad \text{and} \quad - \times (B \times C)$$

without regard to what is in the argument-place of these.

Now, by a "construction," we of course just mean a functor, and by a "relation between constructors" we mean a *morphism of functors* (which is what we are about to define). In the example, it is an isomorphism

$$(- \times B) \times C \;\cong\; - \times (B \times C)$$

of functors $\mathbf{C} \to \mathbf{C}$. In fact, we can of course consider the functors of three arguments:

$$F = (-_1 \times -_2) \times -_3 : \mathbf{C}^3 \to \mathbf{C}$$

and

$$G = -_1 \times (-_2 \times -_3) : \mathbf{C}^3 \to \mathbf{C}$$

and there is an analogous isomorphism

$$F \cong G.$$

But an *isomorphism* is a special morphism, so let us define the general notion first.

Definition 7.6. For categories $\mathbf{C}, \mathbf{D}$ and functors

$$F, G : \mathbf{C} \to \mathbf{D}$$

a *natural transformation* $\vartheta : F \to G$ is a family of arrows in $\mathbf{D}$

$$(\vartheta_C : FC \to GC)_{C \in \mathbf{C}_0}$$

such that, for any $f : C \to C'$ in $\mathbf{C}$, one has $\vartheta_{C'} \circ F(f) = G(f) \circ \vartheta_C$, that is, the following commutes:

$$
\begin{array}{ccc}
FC & \xrightarrow{\;\vartheta_C\;} & GC \\
{\scriptstyle Ff}\downarrow & & \downarrow{\scriptstyle Gf} \\
FC' & \xrightarrow[\;\vartheta_{C'}\;]{} & GC'
\end{array}
$$

Given such a natural transformation $\vartheta : F \to G$, the $\mathbf{D}$-arrow $\vartheta_C : FC \to GC$ is called the *component of* ϑ *at* C.

If you think of a functor $F : \mathbf{C} \to \mathbf{D}$ as a "picture" of $\mathbf{C}$ in $\mathbf{D}$, then you can think of a natural transformation $\vartheta_C : FC \to GC$ as a "cylinder" with such a picture at each end.

7.5 Examples of natural transformations

We have already seen several examples of natural transformations in previous sections, namely the isomorphisms

$$\text{Hom}_{\mathbf{Grp}}(F(1), G) \cong U(G)$$
$$\text{Hom}_{\mathbf{Sets}}(X, \mathbf{2}) \cong \mathcal{P}(X)$$
$$\text{Hom}_{\mathbf{BA}}(B, \mathbf{2}) \cong \text{Ult}(B).$$

There were also the maps from Stone duality,

$$\eta_X : X \to \text{Ult}(\mathcal{P}(X))$$
$$\phi_B : B \to \mathcal{P}(\text{Ult}(B)).$$

We now consider some further examples.

Example 7.7. Consider the free monoid $M(X)$ on a set X and define a natural transformation $\eta : 1_{\mathbf{Sets}} \to UM$, such that each component $\eta_X : X \to UM(X)$ is given by the "insertion of generators" taking every element x to itself, considered as a word.

$$
\begin{array}{ccc}
X & \xrightarrow{\eta_X} & UM(X) \\
\downarrow{\scriptstyle f} & & \downarrow{\scriptstyle UM(f)} \\
Y & \xrightarrow{\eta_Y} & UM(Y)
\end{array}
$$

This is natural, because the homomorphism $M(f)$ on the free monoid $M(X)$ is completely determined by what f does to the generators.

Example 7.8. Let $\mathbf{C}$ be a category with products, and $A \in \mathbf{C}$ fixed. A natural transformation from the functor $A \times - : \mathbf{C} \to \mathbf{C}$ to $1_{\mathbf{C}} : \mathbf{C} \to \mathbf{C}$ is given by taking the component at C to be the second projection

$$\pi_2 : A \times C \to C.$$

From this, together with the pairing operation $\langle -, - \rangle$, one can build up the isomorphism,

$$h : (A \times B) \times C \xrightarrow{\sim} A \times (B \times C).$$

For another such example in more detail, consider the functors

$$\times : \mathbf{C}^2 \to \mathbf{C}$$
$$\bar{\times} : \mathbf{C}^2 \to \mathbf{C}$$

where $\bar{\times}$ is defined on objects by

$$A \mathbin{\bar{\times}} B = B \times A$$

and on arrows by

$$\alpha \mathbin{\bar{\times}} \beta = \beta \times \alpha.$$

Define a "twist" natural transformation $t : \times \to \bar{\times}$ by

$$t_{(A,B)}\langle a, b \rangle = \langle b, a \rangle.$$

To check that the following commutes,

$$
\begin{array}{ccc}
A \times B & \xrightarrow{\;t_{(A,B)}\;} & B \times A \\[2pt]
{\scriptstyle \alpha \times \beta}\Big\downarrow & & \Big\downarrow{\scriptstyle \beta \times \alpha} \\[2pt]
A' \times B' & \xrightarrow[\;t_{(A',B')}\;]{} & B' \times A'
\end{array}
$$

observe that for any generalized elements $a : Z \to A$ and $b : Z \to B$,

$$
\begin{aligned}
(\beta \times \alpha) t_{(A,B)} \langle a, b \rangle &= (\beta \times \alpha) \langle b, a \rangle \\
&= \langle \beta b, \alpha a \rangle \\
&= t_{(A',B')} \langle \alpha a, \beta b \rangle \\
&= t_{(A',B')} \circ (\alpha \times \beta) \langle a, b \rangle.
\end{aligned}
$$

Thus, $t : \times \to \bar{\times}$ is natural. In fact, each component $t_{(A,B)}$ is an isomorphism with inverse $t_{(B,A)}$. This is a simple case of an isomorphism of functors.

Definition 7.9. The *functor category* $\mathrm{Fun}(\mathbf{C}, \mathbf{D})$ has

Objects: functors $F : \mathbf{C} \to \mathbf{D}$,

Arrows: natural transformations $\vartheta : F \to G$.

For each object F, the natural transformation 1_F has components

$$(1_F)_C = 1_{FC} : FC \to FC$$

and the composite natural transformation of $F \xrightarrow{\vartheta} G \xrightarrow{\phi} H$ has components

$$(\phi \circ \vartheta)_C = \phi_C \circ \vartheta_C.$$

Definition 7.10. A *natural isomorphism* is a natural transformation

$$\vartheta : F \to G$$

which is an isomorphism in the functor category $\mathrm{Fun}(\mathbf{C}, \mathbf{D})$.

Lemma 7.11. *A natural transformation $\vartheta : F \to G$ is a natural isomorphism iff each component $\vartheta_C : FC \to GC$ is an isomorphism.*

Proof. Exercise!   □

In our first example, we can therefore say that the isomorphism

$$\vartheta_A : (A \times B) \times C \cong A \times (B \times C)$$

is *natural in* A, meaning that the functors

$$F(A) = (A \times B) \times C$$
$$G(A) = A \times (B \times C)$$

are naturally isomorphic.

Here is a classical example of a natural isomorphism.

Example 7.12. Consider the category

$$\mathrm{Vect}(\mathbb{R})$$

of real vector spaces and linear transformations

$$f : V \to W.$$

Every vector space V has a *dual space*

$$V^* = \mathrm{Vect}(V, \mathbb{R})$$

of linear transformations. And every linear transformation

$$f : V \to W$$

gives rise to a *dual linear transformation*

$$f^* : W^* \to V^*$$

defined by precomposition, $f^*(A) = A \circ f$ for $A : W \to \mathbb{R}$. In brief, $(-)^* = \mathrm{Vect}(-, \mathbb{R}) : \mathbf{Vect}^{\mathrm{op}} \to \mathbf{Vect}$ is the *contravariant representable* functor endowed with vector space structure, just like the examples already considered in Section 7.2.

As in those examples, there is a canonical linear transformation from each vector space to its double dual,

$$\eta_V : V \to V^{**}$$
$$x \mapsto (\mathrm{ev}_x : V^* \to \mathbb{R})$$

where $\mathrm{ev}_x(A) = A(x)$ for every $A : V \to \mathbb{R}$. This map is the component of a natural transformation,

$$\eta : 1_{\mathbf{Vect}} \to {**}$$

since the following always commutes:

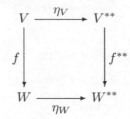

in **Vect**. Indeed, given any $v \in V$ and $A : W \to \mathbb{R}$ in W^*, we have

$$(f^{**} \circ \eta_V)(v)(A) = f^{**}(\text{ev}_v)(A)$$
$$= \text{ev}_v(f^*(A))$$
$$= \text{ev}_v(A \circ f)$$
$$= (A \circ f)(v)$$
$$= A(fv)$$
$$= \text{ev}_{fv}(A)$$
$$= (\eta_W \circ f)(v)(A).$$

Now, it is a well-known fact in linear algebra that every *finite dimensional* vector space V is isomorphic to its dual space $V \cong V^*$ just for reasons of dimension. However, there is no "natural" way to choose such an isomorphism. On the other hand, the natural transformation,

$$\eta_V : V \to V^{**}$$

is a natural isomorphism when V is finite dimensional.

Thus, the formal notion of naturality captures the informal fact that $V \cong V^{**}$ "naturally," unlike $V \cong V^*$.

A similar situation occurs in **Sets**. Here we take 2 instead of $\mathbb{R}$, and the dual A^* of a set A then becomes

$$A^* = \mathcal{P}(A) \cong \textbf{Sets}(A, 2)$$

while the dual of a map $f : A \to B$ is the inverse image $f^* : \mathcal{P}(B) \to \mathcal{P}(A)$.

Note that the exponential evaluation corresponds to (the characteristic function of) the membership relation on $A \times \mathcal{P}(A)$.

$$
\begin{array}{ccc}
2^A \times A & \xrightarrow{\ \epsilon\ } & 2 \\
\cong \downarrow & & \downarrow id \\
A \times \mathcal{P}(A) & \xrightarrow[\tilde{\in}]{} & 2
\end{array}
$$

Transposing again gives a map

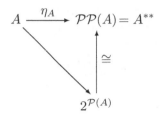

which is described by

$$\eta_A(a) = \{U \subseteq A \mid a \in U\}.$$

In **Sets**, one always has A strictly smaller than $\mathcal{P}(A)$, so $\eta_A : A \to A^{**}$ is *never* an isomorphism. Nonetheless, $\eta : 1_{\textbf{Sets}} \to **$ is a natural transformation, which the reader should prove.

7.6 Exponentials of categories

We now want to show that the category **Cat** of (small) categories and functors is cartesian closed, by showing that any two categories $\textbf{C}, \textbf{D}$ have an exponential $\textbf{D}^{\textbf{C}}$. Of course, we take $\textbf{D}^{\textbf{C}} = \text{Fun}(\textbf{C}, \textbf{D})$, the category of functors and natural transformations, for which we need to prove the required universal mapping property (UMP).

Proposition 7.13. Cat *is cartesian closed, with the exponentials*

$$\textbf{D}^{\textbf{C}} = \text{Fun}(\textbf{C}, \textbf{D}).$$

Before giving the proof, let us note the following. Since exponentials are unique up to isomorphism, this gives us a way to verify that we have found the *"right"* definition of a morphism of functors. For the notion of a natural transformation is completely determined by the requirement that it makes the set $\text{Hom}(\textbf{C}, \textbf{D})$ into an exponential category. This is an example of how category theory can serve as a conceptual tool for discovering new concepts. Before giving the proof, we need the following.

Lemma 7.14 (bifunctor lemma). *Given categories* $\textbf{A}, \textbf{B}$, *and* $\textbf{C}$, *a map of arrows and objects,*

$$F_0 : \textbf{A}_0 \times \textbf{B}_0 \to \textbf{C}_0$$
$$F_1 : \textbf{A}_1 \times \textbf{B}_1 \to \textbf{C}_1$$

is a functor $F : \textbf{A} \times \textbf{B} \to \textbf{C}$ *iff*

1. *F is functorial in each argument:* $F(A, -) : \mathbf{B} \to \mathbf{C}$ *and* $F(-, B) : \mathbf{A} \to \mathbf{C}$
   are functors for all $A \in \mathbf{A}_0$ *and* $B \in \mathbf{B}_0$.

2. *F satisfies the following "interchange law." Given* $\alpha : A \to A' \in \mathbf{A}$ *and*
   $\beta : B \to B' \in \mathbf{B}$, *the following commutes:*

$$
\begin{array}{ccc}
F(A, B) & \xrightarrow{\ F(A, \beta)\ } & F(A, B') \\
{\scriptstyle F(\alpha, B)} \downarrow & & \downarrow {\scriptstyle F(\alpha, B')} \\
F(A', B) & \xrightarrow[\ F(A', \beta)\ ]{} & F(A', B')
\end{array}
$$

   that is, $F(A', \beta) \circ F(\alpha, B) = F(\alpha, B') \circ F(A, \beta)$ *in* $\mathbf{C}$.

Proof. (Lemma) In $\mathbf{A} \times \mathbf{B}$, any arrow

$$
\langle \alpha, \beta \rangle : \langle A, B \rangle \to \langle A', B' \rangle
$$

factors as

$$
\begin{array}{ccc}
\langle A, B \rangle & \xrightarrow{\ \langle 1_A, \beta \rangle\ } & \langle A, B' \rangle \\
{\scriptstyle \langle \alpha, 1_B \rangle} \downarrow & & \downarrow {\scriptstyle \langle \alpha, 1_{B'} \rangle} \\
\langle A', B \rangle & \xrightarrow[\ \langle 1_{A'}, \beta \rangle\ ]{} & \langle A', B' \rangle
\end{array}
$$

So (1) and (2) are clearly necessary. To show that they are also sufficient, we can
define the (proposed) functor:

$$
F(\langle A, B \rangle) = F(A, B)
$$
$$
F(\langle \alpha, \beta \rangle) = F(A', \beta) \circ F(\alpha, B)
$$

The interchange law, together with functoriality in each argument, then ensures
that

$$
F(\alpha', \beta') \circ F(\alpha, \beta) = F(\langle \alpha', \beta' \rangle \circ \langle \alpha, \beta \rangle)
$$

as can be read off from the following diagram:

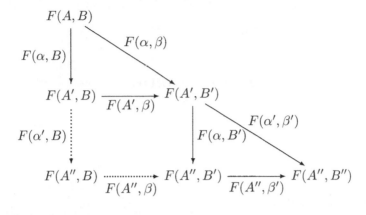

Proof. (Proposition) We need to show:

1. $\epsilon = \mathrm{eval} : \mathrm{Fun}(\mathbf{C}, \mathbf{D}) \times \mathbf{C} \to \mathbf{D}$ is functorial.

2. For any category $\mathbf{X}$ and functor
$$F : \mathbf{X} \times \mathbf{C} \to \mathbf{D}$$
there is a functor
$$\tilde{F} : \mathbf{X} \to \mathrm{Fun}(\mathbf{C}, \mathbf{D})$$
such that $\epsilon \circ (\tilde{F} \times 1_{\mathbf{C}}) = F$.

3. Given any functor
$$G : \mathbf{X} \to \mathrm{Fun}(\mathbf{C}, \mathbf{D}),$$
one has $(\epsilon \circ \widetilde{(G \times 1_{\mathbf{C}})}) = G$.

(1) Using the bifunctor lemma, we show that ϵ is functorial. First, fix $F : \mathbf{C} \to \mathbf{D}$ and consider $\epsilon(F, -) = F : \mathbf{C} \to \mathbf{D}$. This is clearly functorial! Next, fix $C \in \mathbf{C}_0$ and consider $\epsilon(-, C) : \mathrm{Fun}(\mathbf{C}, \mathbf{D}) \to \mathbf{D}$ defined by
$$(\vartheta : F \to G) \mapsto (\vartheta_C : FC \to GC).$$
This is also clearly functorial.

For the interchange law, consider any $\vartheta : F \to G \in \mathrm{Fun}(\mathbf{C}, \mathbf{D})$ and $(f : C \to C') \in \mathbf{C}$, then we need the following to commute:

$$
\begin{array}{ccc}
\epsilon(F, C) & \xrightarrow{\ \vartheta_C\ } & \epsilon(G, C) \\
{\scriptstyle F(f)}\downarrow & & \downarrow{\scriptstyle G(f)} \\
\epsilon(F, C') & \xrightarrow[\ \vartheta_{C'}\ ]{} & \epsilon(G, C')
\end{array}
$$

But this holds because $\epsilon(F, C) = F(C)$ and ϑ is a natural transformation.

The conditions (2) and (3) are now routine. For example, for (2), given

$$F : \mathbf{X} \times \mathbf{C} \to \mathbf{D}$$

let

$$\tilde{F} : \mathbf{X} \to \mathrm{Fun}(\mathbf{C}, \mathbf{D})$$

be defined by

$$\tilde{F}(X)(C) = F(X, C).$$

$\square$

7.7 Functor categories

Let us consider some particular functor categories.

Example 7.15. First, clearly $\mathbf{C}^1 = \mathbf{C}$ for the terminal category $\mathbf{1}$. Next, what about $\mathbf{C}^2$, where $\mathbf{2} = \cdot \to \cdot$ is the single arrow category? This is just the arrow category of $\mathbf{C}$ that we already know,

$$\mathbf{C}^2 = \mathbf{C}^{\to}.$$

Consider instead the discrete category, $2 = \{0, 1\}$. Then clearly,

$$\mathbf{C}^2 \cong \mathbf{C} \times \mathbf{C}.$$

Similarly, for any set I (regarded as a discrete category), we have

$$\mathbf{C}^I \cong \prod_{i \in I} \mathbf{C}.$$

Example 7.16. "Transcendental deduction of natural transformations"
Given the *possibility* of functor categories $\mathbf{D}^{\mathbf{C}}$, we can determine what the objects and arrows therein *must* be as follows:

Objects: these correspond uniquely to functors of the form

$$\mathbf{1} \to \mathbf{D}^{\mathbf{C}}$$

and hence to functors

$$\mathbf{C} \to \mathbf{D}.$$

Arrows: by the foregoing example, arrows in the functor category correspond uniquely to functors of the form

$$\mathbf{1} \to (\mathbf{D}^{\mathbf{C}})^2$$

thus to functors of the form

$$\mathbf{2} \to \mathbf{D}^{\mathbf{C}}$$

and hence to functors

$$\mathbf{C} \times \mathbf{2} \to \mathbf{D}$$

respectively

$$\mathbf{C} \to \mathbf{D}^{\mathbf{2}}.$$

But a functor from $\mathbf{C}$ into the arrow category $\mathbf{D}^{\mathbf{2}}$ (respectively a functor into $\mathbf{D}$ from the cylinder category $\mathbf{C} \times \mathbf{2}$) is exactly a natural transformation between two functors from $\mathbf{C}$ into $\mathbf{D}$, as the reader can see by drawing a picture of the functor's image in $\mathbf{D}$.

Example 7.17. Recall that a (directed) graph can be regarded as a pair of sets and a pair of functions,

$$G_1 \overset{t}{\underset{s}{\rightrightarrows}} G_0$$

where G_1 is the set of edges, G_0 is the set of vertices, and s and t are the source and target operations.

A homomorphism of graphs $h : G \to H$ is a map that preserves sources and targets. In detail, this is a pair of functions $h_1 : G_1 \to H_1$ and $h_0 : G_0 \to H_0$ such that for all edges $e \in G$, we have $sh_1(e) = h_0 s(e)$ and similarly for t as well. But this amounts exactly to saying that the following two diagrams commute:

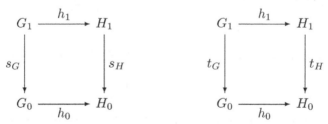

Now consider the category Γ, pictured as follows:

$$\cdot \rightrightarrows \cdot$$

It has exactly two objects and two distinct, parallel, nonidentity arrows. A graph G is then exactly a functor,

$$G : \Gamma \to \mathbf{Sets}$$

and a homomorphism of graphs $h : G \to H$ is exactly a natural transformation between these functors. Thus, the category of graphs is a functor category,

$$\mathbf{Graphs} = \mathbf{Sets}^{\Gamma}.$$

As we see later, it follows from this fact that **Graphs** is cartesian closed.

Example 7.18. Given a product $\mathbf{C} \times \mathbf{D}$ of categories, take the first product projection

$$\mathbf{C} \times \mathbf{D} \to \mathbf{C}$$

and transpose it to get a functor

$$\Delta : \mathbf{C} \to \mathbf{C}^{\mathbf{D}}.$$

For $C \in \mathbf{C}$, the functor $\Delta(C)$ is the "constant C-valued functor,"

- $\Delta(C)(X) = C$ for all $X \in \mathbf{D}_0$
- $\Delta(C)(x) = 1_C$ for all $x \in \mathbf{D}_1$.

Moreover, $\Delta(f) : \Delta(C) \to \Delta(C')$ is the natural transformation, each component of which is f.

Now suppose we have any functor

$$F : \mathbf{D} \to \mathbf{C}$$

and a natural transformation

$$\vartheta : \Delta(C) \to F.$$

Then, the components of ϑ all look like

$$\vartheta_D : C \to F(D)$$

since $\Delta(C)(D) = C$. Moreover, for any $d : D \to D'$ in $\mathbf{D}$, the usual naturality square becomes a triangle, since $\Delta(C)(d) = 1_C$ for all $d : D \to D'$.

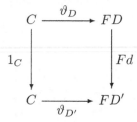

Thus, such a natural transformation $\vartheta : \Delta(C) \to F$ is exactly a cone to the base F (with vertex C). Similarly, a map of cones $\vartheta \to \varphi$ is a constant natural transformation, that is, one of the form $\Delta(h)$ for some $h : C \to D$, making a commutative triangle

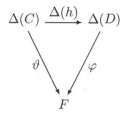

Example 7.19. Take posets P, Q and consider the functor category,

$$Q^P.$$

The functors $P \to Q$, as we know, are just monotone maps, but what is a natural transformation?

$$\vartheta : f \to g$$

For each $p \in P$, we must have

$$\vartheta_p : fp \le gp$$

and if $p \le q$, then there must be a commutative square involving $fp \le fq$ and $gp \le gq$, which, however, is automatic. Thus, the only condition is that $fp \le gp$ for all p, that is, $f \le g$ *pointwise*. Since this is just the usual ordering of the poset Q^P, the exponential poset agrees with the functor category. Thus, we have the following.

Proposition 7.20. *The inclusion functor,*

$$\textbf{Pos} \to \textbf{Cat}$$

preserves CCC structure.

Example 7.21. What happens if we take the functor *category* of two groups G and H?

$$H^G$$

Do we get an exponential of groups? Let us first ask, what is a natural transformation between two group homomorphisms $f, g : G \to H$? Such a map $\vartheta : f \to g$ would be an element $h \in H$ such that for every $x \in G$, we have

$$g(x) \cdot h = h \cdot f(x)$$

or, equivalently,

$$g(x) = h \cdot f(x) \cdot h^{-1}.$$

Therefore, a natural transformation $\vartheta : f \to g$ is an *inner automorphism* $y \mapsto h \cdot y \cdot h^{-1}$ of H (called *conjugation by h*) that takes f to g. Clearly, every such arrow $\vartheta : f \to g$ has an inverse $\vartheta^{-1} : g \to f$ (conjugation by h^{-1}). But H^G is still not usually a group, simply because there may be many *different* homomorphisms $G \to H$, so the functor category H^G has more than one object.

This suggests enlarging the category of groups to include also categories with more than one object, but still having inverses for all arrows. Such categories are called *groupoids*, and have been studied by topologists (they occur as the collection of paths between different points in a topological space). A groupoid can thus be regarded as a generalized group, in which the domains and codomains of elements x and y must match up, as in any category, for the multiplication $x \cdot y$ to be defined.

It is clear that if G and H are any groupoids, then the functor category H^G is also a groupoid. Thus, we have the following proposition, the detailed proof of which is left as an exercise.

Proposition 7.22. *The category* **Grpd** *of groupoids is cartesian closed and the inclusion functor*

$$\textbf{Grpd} \to \textbf{Cat}$$

preserves the CCC structure.

7.8 Monoidal categories

As a further application of natural transformations, we can finally give the general notion of a monoidal category, as opposed to the special case of a *strict* one. Recall from Section 4.1 that a strict monoidal category is by definition a monoid in **Cat**, that is, a category **C** equipped with an associative multiplication functor,

$$\otimes : \textbf{C} \times \textbf{C} \longrightarrow \textbf{C}$$

and a distinguished object I that acts as a unit for $\otimes$. A monoidal category with a discrete category **C** is just a monoid in the usual sense, and every set X gives rise to one of these, with **C** the set of endomorphisms $\text{End}(X)$ under composition. Another example, not discrete, is now had by considering the category $\text{End}(\textbf{D})$ of endo*functors* of an *arbitrary* category **D**, with their natural transformations as arrows; that is, let,

$$\textbf{C} = \text{End}(\textbf{D}), \quad G \otimes F = G \circ F, \quad I = 1_{\textbf{D}}.$$

This can also be seen to be a strict monoidal category. Indeed, the multiplication is clearly associative and has $1_{\textbf{D}}$ as unit, so we just need to check that composition is a bifunctor $\text{End}(\textbf{D}) \times \text{End}(\textbf{D}) \longrightarrow \text{End}(\textbf{D})$. Of course, for this we can use the bifunctor lemma. Fixing F and taking any natural transformation $\alpha : G \to G'$, we have, for any object D,

$$\alpha_{FD} : G(FD) \to G'(FD)$$

which is clearly functorial as an operation $\text{End}(\textbf{D}) \longrightarrow \text{End}(\textbf{D})$. Fixing G and taking $\beta : F \to F'$ gives

$$G(\beta_D) : G(FD) \to G(F'D)$$

which is also easily seen to be functorial. So it just remains to check the exchange law. This comes down to seeing that the square below commutes, which it plainly does just because α is natural.

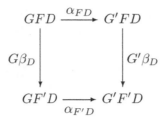

$$GFD \xrightarrow{\alpha_{FD}} G'FD$$

$$G\beta_D \downarrow \qquad \qquad \downarrow G'\beta_D$$

$$GF'D \xrightarrow[\alpha_{F'D}]{} G'F'D$$

Some of the other examples of strict monoidal categories that we have seen involved "product-like" operations such as meets $a \wedge b$ and joins $a \vee b$ in posets. We would like to also capture general products $A \times B$ and coproducts $A + B$ in categories having these; however, these operations are not generally associative on the nose, but only up to isomorphism. Specifically, given any three objects A, B, C in a category with all finite products, we do not have $A \times (B \times C) = (A \times B) \times C$, but instead an isomorphism,

$$A \times (B \times C) \cong (A \times B) \times C.$$

Note, however, that there is exactly one such isomorphism that commutes with all three projections, and it is natural in all three arguments. Similarly, taking a terminal object 1, rather than $1 \times A = A = A \times 1$, we have natural isomorphisms,

$$1 \times A \cong A \cong A \times 1$$

which, again, are uniquely determined by the condition that they commute with the projections. This leads us to the following definition.

Definition 7.23. A *monoidal category* consists of a category $\mathbf{C}$ equipped with a functor

$$\otimes : \mathbf{C} \times \mathbf{C} \longrightarrow \mathbf{C}$$

and a distinguished object I, together with natural isomorphisms,

$$\alpha_{ABC} : A \otimes (B \otimes C) \xrightarrow{\sim} (A \otimes B) \times C,$$

$$\lambda_A : I \otimes A \xrightarrow{\sim} A, \quad \rho_A : A \otimes I \xrightarrow{\sim} A.$$

Moreover, these are required to always make the following diagrams commute:

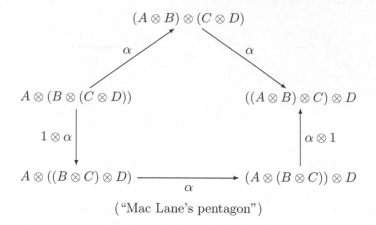

("Mac Lane's pentagon")

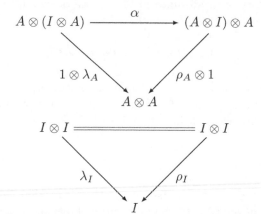

In this precise sense, a monoidal category is thus a category that is strict monoidal "up to natural isomorphism"—where the natural isomorphisms are specified and compatible. An example is, of course, a category with all finite products, where the required equations above are ensured by the UMP of products and the selection of the maps α, λ, ρ as the unique ones commuting with projections. We leave the verification as an exercise. The reader familiar with tensor products of vector spaces, modules, rings, etc., will have no trouble verifying that these, too, give examples of monoidal categories.

A further example comes from an unexpected source: linear logic. The logical operations of linear conjunction and disjunction, sometimes written $P \otimes Q$ and $P \oplus Q$, can be modeled in a monoidal category, usually with extra structure $\sigma_{AB} :$ $A \otimes B \xrightarrow{\sim} B \otimes A$ making these operations "symmetric" (up to isomorphism). Here, too, we leave the verification to the reader familiar with this logical system.

The basic theorem regarding monoidal categories is Mac Lane's coherence theorem, which says that "all diagrams commute." Somewhat more precisely, it says that any diagram in a monoidal category constructed, like those above, just from identities, the functor $\otimes$, and the maps α, λ, ρ will necessarily commute. We shall not state the theorem more precisely than this, nor will we give its somewhat technical proof which, surprisingly, uses ideas from proof theory related to Gentzen's cut elimination theorem! The details can be found in Mac Lane's book, *Categories Work*.

7.9 Equivalence of categories

Before examining some particular functor categories in more detail, we consider one very special application of the concept of natural isomorphism. Consider first the following situation.

Example 7.24. Let $\mathbf{Ord}_{\mathrm{fin}}$ be the category of finite ordinal numbers. Thus, the *objects* are the sets $0, 1, 2, \ldots$, where $0 = \emptyset$ and $n = \{0, \ldots, n-1\}$, while the *arrows* are *all* functions between these sets. Now suppose that for each finite set A we select an ordinal $|A|$ that is its cardinal and an isomorphism,

$$A \cong |A|.$$

Then for each function $f : A \to B$ of finite sets, we have a function $|f|$ by completing the square

$$
\begin{array}{ccc}
A & \xrightarrow{\;\cong\;} & |A| \\
{\scriptstyle f}\downarrow & & \downarrow{\scriptstyle |f|} \\
B & \xrightarrow[\;\cong\;]{} & |B|
\end{array}
\qquad\qquad (7.2)
$$

This clearly gives us a functor

$$|-| : \mathbf{Sets}_{\mathrm{fin}} \to \mathbf{Ord}_{\mathrm{fin}}.$$

Actually, all the maps in the above square are in $\mathbf{Sets}_{\mathrm{fin}}$; so we should also make the inclusion functor

$$i : \mathbf{Ord}_{\mathrm{fin}} \to \mathbf{Sets}_{\mathrm{fin}}$$

explicit. Then we have the selected isos,

$$\vartheta_A : A \xrightarrow{\sim} i|A|$$

and we know by (7.2) that

$$i(|f|) \circ \vartheta_A = \vartheta_B \circ f.$$

This, of course, says that we have a natural isomorphism

$$\vartheta : 1_{\mathbf{Sets_{fin}}} \to i \circ |-|$$

between two functors of the form

$$\mathbf{Sets_{fin}} \to \mathbf{Sets_{fin}}.$$

On the other hand, if we take an ordinal and take its ordinal, we get nothing new,

$$|i(-)| = 1_{\mathbf{Ord_{fin}}} : \mathbf{Ord_{fin}} \to \mathbf{Ord_{fin}}.$$

This is so because, for any finite ordinal n,

$$|i(n)| = n$$

and we can assume that we take $\vartheta_n = 1_n : n \to |i(n)|$, so that also,

$$|i(f)| = f : n \to m.$$

In sum, then, we have a situation where two categories are very similar; but they are *not* the same and they are *not even isomorphic* (why?). This kind of correspondence is what is captured by the notion of equivalence of categories.

Definition 7.25. An *equivalence of categories* consists of a pair of functors

$$E : \mathbf{C} \to \mathbf{D}$$
$$F : \mathbf{D} \to \mathbf{C}$$

and a pair of natural isomorphisms

$$\alpha : 1_{\mathbf{C}} \xrightarrow{\sim} F \circ E \qquad \text{in } \mathbf{C}^{\mathbf{C}}$$
$$\beta : 1_{\mathbf{D}} \xrightarrow{\sim} E \circ F \qquad \text{in } \mathbf{D}^{\mathbf{D}}.$$

In this situation, the functor F is called a *pseudo-inverse* of E. The categories $\mathbf{C}$ and $\mathbf{D}$ are then said to be *equivalent*, written $\mathbf{C} \simeq \mathbf{D}$.

Observe that equivalence of categories is a generalization of isomorphism. Indeed, two categories $\mathbf{C}, \mathbf{D}$ are isomorphic if there are functors.

$$E : \mathbf{C} \to \mathbf{D}$$
$$F : \mathbf{D} \to \mathbf{C}$$

such that

$$1_{\mathbf{C}} = F \circ E$$
$$1_{\mathbf{D}} = E \circ F.$$

In the case of equivalence $\mathbf{C} \simeq \mathbf{D}$, we replace the identity natural transformations by natural isomorphisms. In that sense, equivalence of categories as "isomorphism up to isomorphism."

Experience has shown that the mathematically significant properties of objects are those that are invariant under isomorphisms, and in category theory, identity of objects is a much less important relation than isomorphism. So it is really equivalence of categories that is the more important notion of "similarity" for categories.

In the foregoing example $\mathbf{Sets}_{\text{fin}} \simeq \mathbf{Ord}_{\text{fin}}$, we see that every set is isomorphic to an ordinal, and the maps between ordinals are just the maps between them *as sets*. Thus, we have

1. for every set A, there is some ordinal n with $A \cong i(n)$,
2. for any ordinals n, m, there is an isomorphism,

$$\text{Hom}_{\mathbf{Ord}_{\text{fin}}}(n, m) \cong \text{Hom}_{\mathbf{Sets}_{\text{fin}}}(i(n), i(m))$$

where $i : \mathbf{Ord}_{\text{fin}} \to \mathbf{Sets}_{\text{fin}}$ is the inclusion functor.

In fact, these conditions are characteristic of equivalences, as the following proposition shows.

Proposition 7.26. *The following conditions on a functor* $F : \mathbf{C} \to \mathbf{D}$ *are equivalent:*

1. *F is (part of) an equivalence of categories.*
2. *F is full and faithful and "essentially surjective" on objects: for every $D \in \mathbf{D}$ there is some $C \in \mathbf{C}$ such that $FC \cong D$.*

Proof. (1 implies 2) Take $E : \mathbf{D} \to \mathbf{C}$, and

$$\alpha : 1_{\mathbf{C}} \xrightarrow{\sim} EF$$
$$\beta : 1_{\mathbf{D}} \xrightarrow{\sim} FE.$$

In $\mathbf{C}$, for any C, we then have $\alpha_C : C \xrightarrow{\sim} EF(C)$, and

$$\begin{array}{ccc} C & \xrightarrow{\alpha_C} & EF(C) \\ {\scriptstyle f}\downarrow & & \downarrow{\scriptstyle EF(f)} \\ C' & \xrightarrow[\alpha_{C'}]{} & EF(C') \end{array}$$

commutes for any $f : C \to C'$.

Thus, if $F(f) = F(f')$, then $EF(f) = EF(f')$, so $f = f'$. So F is faithful. Note that by symmetry, E is also faithful.

Now take any arrow

$$h : F(C) \to F(C') \text{ in } \mathbf{D},$$

and consider

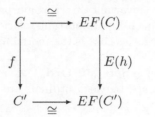

where $f = (\alpha_{C'})^{-1} \circ E(h) \circ \alpha_C$. Then, we have also $F(f) : F(C) \to F(C')$ and

$$EF(f) = E(h) : EF(C) \to EF(C')$$

by the naturality square

$$
\begin{array}{ccc}
C & \xrightarrow{\alpha_C} & EF(C) \\
\downarrow{\scriptstyle f} & & \downarrow{\scriptstyle EF(f)} \\
C' & \xrightarrow{\alpha_{C'}} & EF(C')
\end{array}
$$

Since E is faithful, $F(f) = h$. So F is also *full*.

Finally, for any object $D \in \mathbf{D}$, we have

$$\beta : 1_{\mathbf{D}} \xrightarrow{\sim} FE$$

so

$$\beta_D : D \cong F(ED), \quad \text{for } ED \in \mathbf{C}_0.$$

(2 implies 1) We need to define $E : \mathbf{D} \to \mathbf{C}$ and natural transformations,

$$\alpha : 1_{\mathbf{C}} \xrightarrow{\sim} EF$$

$$\beta : 1_{\mathbf{D}} \xrightarrow{\sim} FE.$$

Since F is essentially surjective, for each $D \in \mathbf{D}_0$, we can *choose* some $E(D) \in \mathbf{C}_0$ along with some $\beta_D : D \xrightarrow{\sim} FE(D)$. That gives E on objects and the proposed components of $\beta : 1_{\mathbf{D}} \to FE$.

Given $h : D \to D'$ in $\mathbf{D}$, consider

$$
\begin{array}{ccc}
D & \xrightarrow{\beta_D} & FE(D) \\
\downarrow{\scriptstyle h} & & \vdots\ {\scriptstyle \beta_{D'} \circ h \circ \beta_D^{-1}} \\
D' & \xrightarrow{\beta_{D'}} & FE(D')
\end{array}
$$

Since $F : \mathbf{C} \to \mathbf{D}$ is full and faithful, there is a unique arrow

$$E(h) : E(D) \to E(D')$$

with $FE(h) = \beta_{D'} \circ h \circ \beta_D^{-1}$. It is easy to see that then $E : \mathbf{D} \to \mathbf{C}$ is a functor and $\beta : 1_{\mathbf{D}} \overset{\sim}{\to} FE$ is clearly a natural isomorphism.

To find $\alpha : 1_{\mathbf{C}} \to EF$, apply F to any C and consider $\beta_{FC} : F(C) \to FEF(C)$. Since F is full and faithful, the preimage of β_{FC} is an isomorphism,

$$\alpha_C = F^{-1}(\beta_{FC}) : C \overset{\sim}{\to} EF(C)$$

which is easily seen to be natural, since β is. $\square$

7.10   Examples of equivalence

Example 7.27. Pointed sets and partial maps
Let **Par** be the category of sets and partial functions. An arrow

$$f : A \rightharpoonup B$$

is a function $|f| : U_f \to B$ for some $U_f \subseteq A$. Identities in **Par** are the same as those in **Sets**, that is, 1_A is the *total* identity function on A. The composite of $f : A \rightharpoonup B$ and $g : B \rightharpoonup C$ is given as follows: Let $U_{(g \circ f)} := f^{-1}(U_g) \subseteq A$, and $|g \circ f| : U_{(g \circ f)} \to C$ is the horizontal composite indicated in the following diagram, in which the square is a pullback:

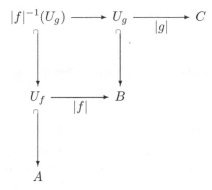

It is easy to see that composition is associative and that the identities are units, so we have a category **Par**.

The category of *pointed sets*,

Sets$_*$

has as objects, sets A equipped with a distinguished "point" $a \in A$, that is, pairs,

$$(A, a) \text{ with } a \in A.$$

Arrows are functions that preserve the point, that is, an arrow $f : (A, a) \to (B, b)$ is a function $f : A \to B$ such that $f(a) = b$.

Now we show:

Proposition 7.28. Par $\simeq$ Sets$_*$

The functors establishing the equivalence are as follows:

$$F : \mathbf{Par} \to \mathbf{Sets}_*$$

is defined on an object A by $F(A) = (A \cup \{*\}, *)$, where $*$ is a new element that we add to A. We also write $A_* = A \cup \{*\}$. For arrows, given $f : A \rightharpoonup B$, $F(f) : A_* \to B_*$ is defined by

$$f_*(x) = \begin{cases} f(x) & \text{if } x \in U_f \\ * & \text{otherwise.} \end{cases}$$

Then clearly $f_*(*_A) = *_B$, so in fact $f_* : A_* \to B_*$ is "pointed," as required.

Coming back, the functor

$$G : \mathbf{Sets}_* \to \mathbf{Par}$$

is defined on an object (A, a) by $G(A, a) = A - \{a\}$ and for an arrow $f : (A, a) \to (B, b)$

$$G(f) : A - \{a\} \rightharpoonup B - \{b\}$$

is the function with domain

$$U_{G(f)} = A - f^{-1}(b)$$

defined by $G(f)(x) = f(x)$ for every $f(x) \neq b$.

Now $G \circ F$ is the identity on **Par**, because we are just adding a new point and then throwing it away. But $F \circ G$ is only naturally isomorphic to $1_{\mathbf{Sets}_*}$, since we have

$$(A, a) \cong ((A - \{a\}) \cup \{*\}, *).$$

These sets are not equal, since $a \neq *$. It still needs to be checked, of course, that F and G are functorial, and that the comparison $(A, a) \cong ((A - \{a\}) \cup \{*\}, *)$ is natural, but we leave these easy verifications to the reader.

Observe that this equivalence implies that **Par** has all limits, since it is equivalent to a category of "algebras" of a very simple type, namely sets equipped with a single, nullary operation, that is, a "constant." We already know that limits of algebras can always be constructed as limits of the underlying sets, and an easy exercise shows that a category equivalent to one with limits of any type also has such limits.

Example 7.29. *Slice categories and indexed families*

For any set I, the functor category $\mathbf{Sets}^I$ is the category of I-indexed sets. The objects are I-indexed families of sets

$$(A_i)_{i \in I}$$

and the arrows are I-indexed families of functions,

$$(f_i : A_i \to B_i)_{i \in I} : (A_i)_{i \in I} \longrightarrow (B_i)_{i \in I}.$$

This category has an equivalent description that is often quite useful: it is equivalent to the slice category of $\mathbf{Sets}$ over I, consisting of arrows $\alpha : A \to I$ and "commutative triangles" over I (see Section 1.6),

$$\mathbf{Sets}^I \simeq \mathbf{Sets}/I.$$

Indeed, define functors

$$\Phi : \mathbf{Sets}^I \longrightarrow \mathbf{Sets}/I$$

$$\Psi : \mathbf{Sets}/I \longrightarrow \mathbf{Sets}^I$$

on objects as follows:

$$\Phi((A_i)_{i \in I}) = \pi : \coprod_{i \in I} A_i \to I \quad \text{(the indexing projection)},$$

where the coproduct is conveniently taken to be

$$\coprod_{i \in I} A_i = \{(i, a) \,|\, a \in A_i\}.$$

And coming back, we have

$$\Psi(\alpha : A \to I) = (\alpha^{-1}\{i\})_{i \in I}.$$

The effect on arrows is analogous and easily inferred. We leave it as an exercise to show that these are indeed mutually pseudo-inverse functors. (Why are they *not* inverses?)

The equivalent description of $\mathbf{Sets}^I$ as $\mathbf{Sets}/I$ leads to the idea that, for a general category $\mathcal{E}$, the slice category $\mathcal{E}/X$, for any object X, can also be regarded as the category of "X-indexed objects of $\mathcal{E}$", although the functor category $\mathcal{E}^X$ usually does not make sense. This provides a precise notion of an "X-indexed family of objects E_x of $\mathcal{E}$," namely as a map $E \to X$.

For instance, in topology, there is the notion of a "fiber bundle" as a continuous function $\pi : Y \to X$, thought of as a family of spaces $Y_x = \pi^{-1}(x)$, the "fibers" of π, varying continuously in a parameter $x \in X$. Similarly, in dependent type theory there are "dependent types" $x : X \vdash A(x)$, thought of as families of types indexed over a type. These can be modeled as objects $[\![A]\!] \to [\![X]\!]$ in the slice category $\mathcal{E}/[\![X]\!]$ over the interpretation of the (closed) type X as an object of a category $\mathcal{E}$.

If $\mathcal{E}$ has pullbacks, reindexing of an "indexed family" along an arrow $f : Y \to X$ in $\mathcal{E}$ is represented by the pullback functor $f^* : \mathcal{E}/X \to \mathcal{E}/Y$. This is motivated by the fact that in **Sets** the following diagram commutes (up to natural isomorphism) for any $f : J \to I$:

$$
\begin{array}{ccc}
\mathbf{Sets}^I & \xrightarrow{\;\simeq\;} & \mathbf{Sets}/I \\
{\scriptstyle \mathbf{Sets}^f}\big\downarrow & & \big\downarrow{\scriptstyle f^*} \\
\mathbf{Sets}^J & \xrightarrow[\simeq]{\;\;\;} & \mathbf{Sets}/J
\end{array}
$$

where the functor $\mathbf{Sets}^f$ is the reindexing along f:

$$(\mathbf{Sets}^f(A_i))_j = A_{f(j)}.$$

Moreover, there are also functors going in the other direction,

$$\Sigma_f, \Pi_f : \mathbf{Sets}/J \longrightarrow \mathbf{Sets}/I$$

which, in terms of indexed families, are given by taking sums and products of the fibers:

$$(\Sigma_f(A_j))_i = \sum_{f(j)=i} A_j$$

and similarly for Π. These functors can be characterized in terms of the pullback functor f^* (as adjoints, see Section 9.7), and so also make sense in categories more general than **Sets**, where there are no "indexed families" in the usual sense. For instance, in dependent type theory, these operations are formalized by logical rules of inference similar to those for the existential and universal quantifier, and the resulting category of types has such operations of dependent sums and products.

Example 7.30. *Stone duality*
Many examples of equivalences of categories are given by what are called "dualities." Often, classical duality theorems are not of the form $\mathbf{C} \cong \mathbf{D}^{\mathrm{op}}$ (much less $\mathbf{C} = \mathbf{D}^{\mathrm{op}}$), but rather $\mathbf{C} \simeq \mathbf{D}^{\mathrm{op}}$, that is, $\mathbf{C}$ is equivalent to the opposite (or "dual") category of $\mathbf{D}$. This is because the duality is established by a construction that returns the original thing only up to isomorphism, not "on the nose." Here is a simple example, which is a very special case of the far-reaching *Stone duality* theorem.

Proposition 7.31. *The category of finite Boolean algebras is equivalent to the opposite of the category of finite sets,*

$$\mathbf{BA}_{\mathrm{fin}} \simeq \mathbf{Sets}_{\mathrm{fin}}^{\mathrm{op}}.$$

Proof. The functors involved here are the contravariant powerset functor

$$\mathcal{P}^{\mathbf{BA}} : \mathbf{Sets}_{\text{fin}}^{\text{op}} \to \mathbf{BA}_{\text{fin}}$$

on one side (the powerset of a finite set is finite!). Going back, we use the functor,

$$A : \mathbf{BA}_{\text{fin}}^{\text{op}} \to \mathbf{Sets}_{\text{fin}}$$

taking the set of *atoms* of a Boolean algebra,

$$A(\mathcal{B}) = \{a \in \mathcal{B} \mid 0 < a \text{ and } (b < a \text{ implies } b = 0)\}.$$

In the finite case, this is isomorphic to the ultrafilter functor that we have already studied (see Section 7.3).

Lemma 7.32. *For any finite Boolean algebra $\mathcal{B}$, there is an isomorphism between atoms a in $\mathcal{B}$ and ultrafilters $U \subseteq \mathcal{B}$, given by*

$$U \mapsto \bigwedge_{b \in U} b$$

and

$$a \mapsto \uparrow(a).$$

Proof. If a is an atom, then $\uparrow(a)$ is an ultrafilter, since for any b either $a \wedge b = a$ and then $b \in \uparrow(a)$ or $a \wedge b = 0$ and so $\neg b \in \uparrow(a)$.

If $U \subseteq \mathcal{B}$ is an ultrafilter then $0 < \bigwedge_{b \in U} b$, because, since U is finite and closed under intersections, we must have $\bigwedge_{b \in U} b \in U$. If $0 \neq b_0 < \bigwedge_{b \in U} b$ then b_0 is not in U, so $\neg b_0 \in U$. But then $b_0 < \neg b_0$ and so $b_0 = 0$.

Plainly, $U \subseteq \uparrow(\bigwedge_{b \in U} b)$ since $b \in U$ implies $\bigwedge_{b \in U} b \subseteq b$. Now let $\bigwedge_{b \in U} b \leq a$ for some a not in U. Then, $\neg a \in U$ implies that also $\bigwedge_{b \in U} b \leq \neg a$, and so $\bigwedge_{b \in U} b \leq a \wedge \neg a = 0$, which is impossible. $\qquad\square$

Since we know that the set of ultrafilters $\text{Ult}(\mathcal{B})$ is contravariantly functorial (it is represented by the Boolean algebra **2**, see Section 7.3), we therefore also have a contravariant functor of atoms $A \cong \text{Ult}$. The explicit description of this functor is this: if $h : \mathcal{B} \to \mathcal{B}'$ and $a' \in A(\mathcal{B}')$, then it follows from the lemma that there is a *unique* atom $a \in \mathcal{B}$ such that $a' \leq h(b)$ iff $a \leq b$ for all $b \in \mathcal{B}$. To find this atom a, take the intersection over the ultrafilter $h^{-1}(\uparrow(a'))$,

$$A(a') = a = \bigwedge_{a' \leq h(b)} b.$$

Thus, we get a function

$$A(h) : A(\mathcal{B}') \to A(\mathcal{B}).$$

Of course, we must still check that this is a pseudo-inverse for $\mathcal{P}^{\mathbf{BA}} : \mathbf{Sets}_{\text{fin}}^{\text{op}} \to$ $\mathbf{BA}_{\text{fin}}$. The required natural isomorphisms,

$$\alpha_X : X \to A(\mathcal{P}(X))$$

$$\beta_{\mathcal{B}} : \mathcal{B} \to \mathcal{P}(A(\mathcal{B}))$$

are explicitly described as follows:

The atoms in a finite powerset $\mathcal{P}(X)$ are just the singletons $\{x\}$ for $x \in X$, thus $\alpha_X(x) = \{x\}$ is clearly an isomorphism.

To define $\beta_{\mathcal{B}}$, let

$$\beta_{\mathcal{B}}(b) = \{a \in A(\mathcal{B}) \mid a \leq b\}.$$

To see that $\beta_{\mathcal{B}}$ is also iso, consider the proposed inverse,

$$(\beta_{\mathcal{B}})^{-1}(B) = \bigvee_{a \in B} a \qquad \text{for } B \subseteq A(\mathcal{B}).$$

The isomorphism then follows from the following lemma, the proof of which is routine.

Lemma 7.33. *For any finite Boolean algebra $\mathcal{B}$,*

1. $b = \bigvee\{a \in A(\mathcal{B}) \mid a \leq b\}$.

2. If a is an atom and $a \leq b \vee b'$, then $a \leq b$ or $a \leq b'$.

Of course, one must still check that α and β really are natural transformations. This is left to the reader. $\qquad\qquad\qquad\qquad\qquad\qquad\qquad\qquad\quad\square$

Finally, we remark that the duality

$$\mathbf{BA}_{\text{fin}} \simeq \mathbf{Sets}_{\text{fin}}^{\text{op}}$$

extends to one between *all* sets on the one side and the complete, atomic Boolean algebras, on the other,

$$\mathbf{caBA} \simeq \mathbf{Sets}^{\text{op}},$$

where a Boolean algebra $\mathcal{B}$ is *complete* if every subset $U \subseteq \mathcal{B}$ has a join $\bigvee U \in \mathcal{B}$ and a *complete homomorphism* preserves these joins and $\mathcal{B}$ is *atomic* if every nonzero element $0 \neq b \in \mathcal{B}$ has some $a \leq b$ with a an atom.

Moreover, this is just the discrete case of the full Stone duality theorem, which states an equivalence between the category of *all* Boolean algebras and the opposite of a certain category of topological spaces, called "Stone spaces," and all continuous maps between them. For details, see Johnstone (1982).

7.11   Exercises

1. Consider the (covariant) composite functor,

$$\mathcal{F} = \mathcal{P}^{\mathbf{BA}} \circ \mathrm{Ult}^{\mathrm{op}} : \mathbf{BA} \to \mathbf{Sets}^{\mathrm{op}} \to \mathbf{BA}$$

taking each Boolean algebra B to the powerset algebra of sets of ultrafilters in B. Note that

$$\mathcal{F}(B) \cong \mathrm{Hom}_{\mathbf{Sets}}(\mathrm{Hom}_{\mathbf{BA}}(B, \mathbf{2}), \mathbf{2})$$

is a sort of "double-dual" Boolean algebra. There is always a homomorphism,

$$\phi_B : B \to \mathcal{F}(B)$$

given by $\phi_B(b) = \{\mathcal{V} \in \mathrm{Ult}(B) \mid b \in \mathcal{V}\}$. Show that for any Boolean homomorphism $h : A \to B$, the following square commutes:

$$
\begin{array}{ccc}
A & \xrightarrow{\phi_A} & \mathcal{F}(A) \\
\downarrow{\scriptstyle h} & & \downarrow{\scriptstyle \mathcal{F}(h)} \\
B & \xrightarrow[\phi_B]{} & \mathcal{F}(B)
\end{array}
$$

2. Show that the homomorphism $\phi_B : B \to \mathcal{F}(B)$ in the foregoing problem is always injective (use the Boolean prime ideal theorem). This is the classical "Stone representation theorem," stating that every Boolean algebra is isomorphic to a "field of sets," that is, a sub-Boolean algebra of a powerset. Is the functor $\mathcal{F}$ faithful?

3. Prove that for any *finite* Boolean algebra B, the "Stone representation"

$$\phi : B \to \mathcal{P}(\mathrm{Ult}(B))$$

is in fact an isomorphism of Boolean algebras. (Note the similarity to the case of finite dimensional vector spaces.) This concludes the proof that we have an equivalence of categories,

$$\mathbf{BA}_{\mathrm{fin}} \simeq \mathbf{Sets}^{\mathrm{op}}_{\mathrm{fin}}$$

This is the "finite" case of Stone duality.

4. Consider the forgetful functors

$$\mathbf{Groups} \xrightarrow{U} \mathbf{Monoids} \xrightarrow{V} \mathbf{Sets}$$

Say whether each is faithful, full, injective on arrows, surjective on arrows, injective on objects, and surjective on objects.

5. Make every poset $(X, \leq)$ into a topological space by letting $U \subseteq X$ be open just if $x \in U$ and $x \leq y$ implies $y \in U$ (U is "closed upward"). This is called the *Alexandroff topology* on X. Show that it gives a functor

$$A : \mathbf{Pos} \to \mathbf{Top}$$

from posets and monotone maps to spaces and continuous maps by showing that any monotone map of posets $f : P \to Q$ is continuous with respect to this topology on P and Q (the inverse image of an open set must be open).

Is A faithful? Is it full?

6. Prove that every functor $F : \mathbf{C} \to \mathbf{D}$ can be factored as $D \circ E = F$,

$$\mathbf{C} \xrightarrow{E} \mathbf{E} \xrightarrow{D} \mathbf{D}$$

in the following two ways:

(a) $E : \mathbf{C} \to \mathbf{E}$ is bijective on objects and full, and $D : \mathbf{E} \to \mathbf{D}$ is faithful;

(b) $E : \mathbf{C} \to \mathbf{E}$ surjective on objects and $D : \mathbf{E} \to \mathbf{D}$ is injective on objects and full and faithful.

When do the two factorizations agree?

7. Show that a natural transformation is a natural isomorphism just if each of its components is an isomorphism. Is the same true for monomorphisms?

8. Show that a functor category $\mathbf{D}^{\mathbf{C}}$ has binary products if $\mathbf{D}$ does (construct the product of two functors F and G "objectwise": $(F \times G)(C) = F(C) \times G(C)$).

9. Show that the map of sets

$$\eta_A : A \longrightarrow \mathcal{P}\mathcal{P}(A)$$
$$a \longmapsto \{U \subseteq A | a \in U\}$$

is the component at A of a natural transformation $\eta : 1_{\mathbf{Sets}} \to \mathcal{P}\mathcal{P}$, where $\mathcal{P} : \mathbf{Sets}^{\mathrm{op}} \to \mathbf{Sets}$ is the (contravariant) powerset functor.

10. Let $\mathbf{C}$ be a locally small category. Show that there is a functor

$$\mathrm{Hom} : \mathbf{C}^{\mathrm{op}} \times \mathbf{C} \to \mathbf{Sets}$$

such that for each object C of $\mathbf{C}$,

$$\mathrm{Hom}(C, -) : \mathbf{C} \to \mathbf{Sets}$$

is the covariant representable functor and

$$\mathrm{Hom}(-, C) : \mathbf{C}^{\mathrm{op}} \to \mathbf{Sets}$$

is the contravariant one. (Hint: *use the bifunctor lemma.*)

11. Recall from the text that a *groupoid* is a category in which every arrow is an isomorphism. Prove that the category of groupoids is cartesian closed.

12. Let $\mathbf{C} \cong \mathbf{D}$ be equivalent categories. Show that $\mathbf{C}$ has binary products if and only if $\mathbf{D}$ does.

13. What sorts of properties of categories do *not* respect equivalence? Find one that respects isomorphism, but not equivalence.

14. Complete the proof that $\mathbf{Par} \cong \mathbf{Sets}_*$.

15. Show that equivalence of categories is an equivalence relation.

16. A category is *skeletal* if isomorphic objects are always identical. Show that every category is equivalent to a skeletal subcategory. (Every category has a "skeleton.")

17. Complete the proof that, for any set I, the category of I-indexed families of sets, regarded as the functor category $\mathbf{Sets}^I$, is equivalent to the slice category $\mathbf{Sets}/I$ of sets over I,

$$\mathbf{Sets}^I \simeq \mathbf{Sets}/I.$$

Show that reindexing of families along a function $f : J \to I$, given by precomposition,

$$\mathbf{Sets}^f((A_i)_{i \in I}) = (A_{f(j)})_{j \in J}$$

is represented by pullback, in the sense that the following diagram of categories and functors commutes up to natural isomorphism:

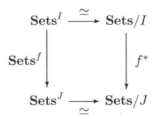

Here $f^* : \mathbf{Sets}/J \to \mathbf{Sets}/I$ is the pullback functor along $f : J \to I$. Finally, infer that $\mathbf{Sets}/2 \simeq \mathbf{Sets} \times \mathbf{Sets}$, and similarly for any n other than 2.

18. Show that a category with finite products is a monoidal category. Infer that the same is true for any category with finite coproducts.

8

CATEGORIES OF DIAGRAMS

In this chapter, we prove a very useful technical result called the Yoneda lemma, and then employ it in the study of the important categories of set-valued functors or "diagrams." The Yoneda lemma is perhaps the single most used result in category theory. It can be seen as a straightforward generalization of some simple facts about monoids and posets, yet it has much more far-reaching applications.

8.1 Set-valued functor categories

We are going to focus on special functor categories of the form

$$\mathbf{Sets}^{\mathbf{C}}$$

where the category $\mathbf{C}$ is locally small. Thus, the objects are set-valued functors,

$$F, G : \mathbf{C} \to \mathbf{Sets}$$

(sometimes called "diagrams on $\mathbf{C}$"), and the arrows are natural transformations

$$\alpha, \beta : F \to G.$$

Where $\mathbf{C} = \mathrm{P}$, a poset, we have already considered such functors as "variable sets," that is, sets F_i depending on a parameter $i \in \mathrm{P}$. The general case of a non-poset $\mathbf{C}$ similarly admits an interpretation as "variable sets": such a functor F gives a family of sets FC and transitions $FC \to FC'$ showing how the sets change according to every $C \to C'$. For instance, $\mathbf{C}$ might be the category $\mathbf{Sets}_{\mathrm{fin}}$ of all finite sets (of finite sets, ...) and all functions between them. Then in $\mathbf{Sets}^{\mathbf{Sets}_{\mathrm{fin}}}$ there is for example the inclusion functor $U : \mathbf{Sets}_{\mathrm{fin}} \to \mathbf{Sets}$, which can be regarded as a "generic" or variable finite set, along with the functors $U \times U$, $U + U$, etc., which are "variable" structures of these kinds.

Given any such category $\mathbf{Sets}^{\mathbf{C}}$, remember that we can evaluate any commutative diagram,

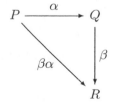

at any object C to get a commutative diagram in **Sets**,

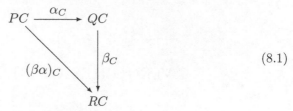

$$(8.1)$$

Thus, for each object C, there is an evaluation functor

$$\mathrm{ev}_C : \mathbf{Sets}^{\mathbf{C}} \to \mathbf{Sets}.$$

Moreover, naturality means that if we have any arrow $f : D \to C$, we get a "cylinder" over the diagram (8.1) in **Sets**.

Another way of thinking about such functor categories that was already considered in Section 7.7 is suggested by considering the case where **C** is the category Γ pictured as

$$1 \rightrightarrows 0$$

Then a set-valued functor $G : \Gamma \to \mathbf{Sets}$ is just a graph, and a natural transformation $\alpha : G \to H$ is a graph homomorphism. Thus, for this case,

$$\mathbf{Sets}^{\Gamma} = \mathbf{Graphs}.$$

This suggests regarding an arbitrary category of the form $\mathbf{Sets}^{\mathbf{C}}$ as a generalized "category of structured sets" and their "homomorphisms"; indeed, this is a very useful way of thinking of such functors and their natural transformations.

Another basic example is the category $\mathbf{Sets}^{\Delta^{\mathrm{op}}}$, where the index category Δ is the category of finite ordinals that we already met in Chapter 7. The objects of $\mathbf{Sets}^{\Delta^{\mathrm{op}}}$ are called *simplicial sets*, and are used in topology to compute the homology, cohomology, and homotopy of spaces. Since Δ looks like

$$0 \longrightarrow 1 \rightrightarrows 2 \Rrightarrow 3 \quad \cdots$$

(satisfying the simplicial identities), a simplicial set $S : \Delta^{\mathrm{op}} \to \mathbf{Sets}$ looks like this:

$$S_0 \longleftarrow S_1 \rightrightarrows S_2 \Rrightarrow S_3 \quad \cdots$$

(satisfying the corresponding identities). For example, one can take $S_n = S^n = S \times \ldots \times S$ (n times) for a fixed set S to get a (rather trivial) simplicial set, with the maps being the evident product projections and generalized diagonals. More interestingly, for a fixed poset P, one takes

$$S(\mathrm{P})_n = \{(p_1, \ldots, p_n) \in \mathrm{P}^n \mid p_1 \leq \ldots \leq p_n\},$$

with the evident projections and inclusions; this is called the "simplicial nerve" of the poset P.

8.2 The Yoneda embedding

Among the objects of $\mathbf{Sets}^{\mathbf{C}}$ are certain very special ones, namely the (covariant) representable functors,

$$\mathrm{Hom}_{\mathbf{C}}(C, -) : \mathbf{C} \to \mathbf{Sets}.$$

Observe that for each $h : C \to D$ in $\mathbf{C}$, we have a natural transformation

$$\mathrm{Hom}_{\mathbf{C}}(h, -) : \mathrm{Hom}_{\mathbf{C}}(D, -) \to \mathrm{Hom}_{\mathbf{C}}(C, -)$$

(note the direction!) where the component at X is defined by precomposition:

$$(f : D \to X) \mapsto (f \circ h : C \to X).$$

Thus, we have a *contravariant* functor

$$k : \mathbf{C}^{\mathrm{op}} \to \mathbf{Sets}^{\mathbf{C}}$$

defined by $k(C) = \mathrm{Hom}_{\mathbf{C}}(C, -)$. Of course, this functor k is just the exponential transpose of the bifunctor

$$\mathrm{Hom}_{\mathbf{C}} : \mathbf{C}^{\mathrm{op}} \times \mathbf{C} \to \mathbf{Sets}$$

which was shown as an exercise to be functorial.

If we instead transpose $\mathrm{Hom}_{\mathbf{C}}$ with respect to its other argument, we get a *covariant* functor,

$$y : \mathbf{C} \to \mathbf{Sets}^{\mathbf{C}^{\mathrm{op}}}$$

from $\mathbf{C}$ to a category of *contravariant* set-valued functors, sometimes called "presheaves." (Or, what amounts to the same thing, we can put $\mathbf{D} = \mathbf{C}^{\mathrm{op}}$ and apply the previous considerations to $\mathbf{D}$ in place of $\mathbf{C}$.) More formally:

Definition 8.1. The *Yoneda embedding* is the functor $y : \mathbf{C} \to \mathbf{Sets}^{\mathbf{C}^{\mathrm{op}}}$ taking $C \in \mathbf{C}$ to the contravariant representable functor,

$$yC = \mathrm{Hom}_{\mathbf{C}}(-, C) : \mathbf{C}^{\mathrm{op}} \to \mathbf{Sets}$$

and taking $f : C \to D$ to the natural transformation,

$$yf = \mathrm{Hom}_{\mathbf{C}}(-, f) : \mathrm{Hom}_{\mathbf{C}}(-, C) \to \mathrm{Hom}_{\mathbf{C}}(-, D).$$

A functor $F : \mathbf{C} \to \mathbf{D}$ is called an *embedding* if it is full, faithful, and injective on objects. We soon show that y really *is* an embedding; this is a corollary of the Yoneda lemma.

One should thus think of the Yoneda embedding y as a "representation" of $\mathbf{C}$ in a category of set-valued functors and natural transformations on *some* index category. Compared to the Cayley representation considered in Section 1.5, this has the virtue of being *full*: any map $\vartheta : yC \to yD$ in $\mathbf{Sets}^{\mathbf{C}^{\mathrm{op}}}$ comes from a unique map $h : C \to D$ in $\mathbf{C}$ as $yh = \vartheta$. Indeed, recall that the Cayley representation of a group G was an injective group homomorphism

$$G \rightarrowtail \mathrm{Aut}(|G|) \subseteq |G|^{|G|}$$

where each $g \in G$ is represented as an automorphism $\tilde{g}$ of the set $|G|$ of elements (i.e., a "permutation"), by letting it "act on the left,"

$$\tilde{g}(x) = g \cdot x$$

and the group multiplication is represented by composition of permutations,

$$\widetilde{g \cdot h} = \tilde{g} \circ \tilde{h}.$$

We also showed a generalization of this representation to arbitrary categories. Thus for any monoid M, there is an analogous representation

$$M \rightarrowtail \mathrm{End}(|M|) \subseteq |M|^{|M|}$$

by left action, representing the elements of M as endomorphisms of $|M|$.

Similarly, any poset P can be represented as a poset of subsets and inclusions by considering the poset $\mathrm{Low}(P)$ of "lower sets" $A \subseteq P$, that is, subsets that are "closed down" in the sense that $a' \leq a \in A$ implies $a' \in A$, ordered by inclusion. Taking the "principal lower set"

$$\downarrow(p) = \{q \in P \mid q \leq p\}$$

of each element $p \in P$ determines a monotone injection

$$\downarrow : P \rightarrowtail \mathrm{Low}(P) \subseteq \mathcal{P}(|P|)$$

such that $p \leq q$ iff $\downarrow(p) \subseteq \downarrow(q)$.

The representation given by the Yoneda embedding is closely related to these, but "better" in that it cuts down the arrows in the codomain category to just those in the image of the representation functor $y : \mathbf{C} \to \mathbf{Sets}^{\mathbf{C}^{\mathrm{op}}}$ (since y is full). Indeed, there may be many automorphisms $\alpha : G \to G$ of a group G that are not left actions by an element, but if we require α to commute with all *right* actions $\alpha(x \cdot g) = \alpha(x) \cdot g$, then α must itself be a left action. This is what the Yoneda embedding does in general; it adds enough "structure" to the objects yA in the image of the representation that the only "homomorphisms" $\vartheta : yA \to yB$ between those objects are the representable ones $\vartheta = yh$ for some $h : A \to B$. In this sense, the Yoneda embedding y represents the objects and arrows of $\mathbf{C}$ as certain "structured sets" and (*all of*) their "homomorphisms."

8.3 The Yoneda lemma

Lemma 8.2 (Yoneda). *Let $\mathbf{C}$ be locally small. For any object $C \in \mathbf{C}$ and functor $F \in \mathbf{Sets},^{\mathbf{C}^{\mathrm{op}}}$ there is an isomorphism*

$$\mathrm{Hom}(yC, F) \cong FC$$

which, moreover, is natural in both F and C.

Here

(1) the Hom is $\mathrm{Hom}_{\mathbf{Sets}^{\mathbf{C}^{\mathrm{op}}}}$,

(2) naturality in F means that, given any $\vartheta : F \to G$, the following diagram commutes:

$$
\begin{array}{ccc}
\mathrm{Hom}(yC, F) & \xrightarrow{\;\cong\;} & FC \\
{\scriptstyle \mathrm{Hom}(yC, \vartheta)} \downarrow & & \downarrow {\scriptstyle \vartheta_C} \\
\mathrm{Hom}(yC, G) & \xrightarrow[\;\cong\;]{} & GC
\end{array}
$$

(3) naturality in C means that, given any $h : C \to D$, the following diagram commutes:

$$
\begin{array}{ccc}
\mathrm{Hom}(yC, F) & \xrightarrow{\;\cong\;} & FC \\
{\scriptstyle \mathrm{Hom}(yh, F)} \uparrow & & \uparrow {\scriptstyle Fh} \\
\mathrm{Hom}(yD, F) & \xrightarrow[\;\cong\;]{} & FD
\end{array}
$$

Proof. To define the desired isomorphism,

$$\eta_{C,F} : \mathrm{Hom}(yC, F) \xrightarrow{\;\cong\;} FC$$

take $\vartheta : yC \to F$ and let

$$\eta_{C,F}(\vartheta) = \vartheta_C(1_C)$$

which we also write as

$$x_\vartheta = \vartheta_C(1_C) \tag{8.2}$$

where $\vartheta_C : \mathbf{C}(C,C) \to FC$ and so $\vartheta_C(1_C) \in FC$.

Conversely, given any $a \in FC$, we define the natural transformation $\vartheta_a : yC \to F$ as follows. Given any C', we define the component

$$(\vartheta_a)_{C'} : \mathrm{Hom}(C', C) \to FC'$$

by setting

$$(\vartheta_a)_{C'}(h) = F(h)(a) \tag{8.3}$$

for $h : C' \to C$.

To show that ϑ_a is natural, take any $f : C'' \to C'$, and consider the following diagram:

$$
\begin{array}{ccc}
\mathrm{Hom}(C'',C) & \xrightarrow{\;(\vartheta_a)_{C''}\;} & FC'' \\[2pt]
\big\uparrow{\scriptstyle \mathrm{Hom}(f,C)} & & \big\uparrow{\scriptstyle F(f)} \\[2pt]
\mathrm{Hom}(C',C) & \xrightarrow[\;(\vartheta_a)_{C'}\;]{} & FC'
\end{array}
$$

We then calculate, for any $h \in yC(C')$

$$
\begin{aligned}
(\vartheta_a)_{C''} \circ \mathrm{Hom}(f,C)(h) &= (\vartheta_a)_{C''}(h \circ f) \\
&= F(h \circ f)(a) \\
&= F(f) \circ F(h)(a) \\
&= F(f)(\vartheta_a)_{C'}(h).
\end{aligned}
$$

So ϑ_a is indeed natural.

Now to show that ϑ_a and x_ϑ are mutually inverse, let us calculate ϑ_{x_ϑ} for a given $\vartheta : yC \to F$. First, just from the definitions (8.2) and (8.3), we have that for any $h : C' \to C$,

$$
(\vartheta_{(x_\vartheta)})_{C'}(h) = F(h)(\vartheta_C(1_C)).
$$

But since ϑ is natural, the following commutes:

$$
\begin{array}{ccc}
yC(C) & \xrightarrow{\;\vartheta_C\;} & FC \\[2pt]
\big\downarrow{\scriptstyle yC(h)} & & \big\downarrow{\scriptstyle Fh} \\[2pt]
yC(C') & \xrightarrow[\;\vartheta_{C'}\;]{} & FC'
\end{array}
$$

So, continuing,

$$
\begin{aligned}
(\vartheta_{(x_\vartheta)})_{C'}(h) &= F(h)(\vartheta_C(1_C)) \\
&= \vartheta_{C'} \circ yC(h)(1_C) \\
&= \vartheta_{C'}(h).
\end{aligned}
$$

Therefore, $\vartheta_{(x_\vartheta)} = \vartheta$.

Going the other way around, for any $a \in FC$, we have

$$x_{\vartheta_a} = (\vartheta_a)_C(1_C)$$
$$= F(1_C)(a)$$
$$= 1_{FC}(a)$$
$$= a.$$

Thus, $\mathrm{Hom}(yC, F) \cong FC$, as required.

The naturality claims are also easy: given $\phi : F \to F'$, taking $\vartheta \in \mathrm{Hom}(yC, F)$, and chasing around the diagram

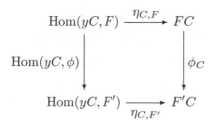

we get

$$\phi_C(x_\vartheta) = \phi_C(\vartheta_C(1_C))$$
$$= (\phi\vartheta)_C(1_C)$$
$$= x_{(\phi\vartheta)}$$
$$= \eta_{C,F'}(\mathrm{Hom}(yC, \phi)(\vartheta)).$$

For naturality in C, take some $f : C' \to C$. We then have

$$\eta_{C'}(yf)^*(\vartheta) = \eta_{C'}(\vartheta \circ yf)$$
$$= (\vartheta \circ yf)_{C'}(1_{C'})$$
$$= \vartheta_{C'} \circ (yf)_{C'}(1_{C'})$$
$$= \vartheta_{C'}(f \circ 1_{C'})$$
$$= \vartheta_{C'}(f)$$
$$= \vartheta_{C'}(1_C \circ f)$$
$$= \vartheta_{C'} \circ (yC)(f)(1_C)$$
$$= F(f) \circ \vartheta_C(1_C)$$
$$= F(f)\eta_C(\vartheta).$$

The penultimate equation is by the naturality square:

$$
\begin{array}{ccc}
yC(C) & \xrightarrow{\ \vartheta_C\ } & F(C) \\
{\scriptstyle yC(f)}\Big\downarrow & & \Big\downarrow{\scriptstyle F(f)} \\
yC(C') & \xrightarrow[\ \vartheta_{C'}\ ]{} & F(C')
\end{array}
$$

Therefore, $\eta_{C'} \circ (yf)^* = F(f) \circ \eta_C$. $\qquad\qquad\qquad\qquad\qquad\qquad\square$

The Yoneda lemma is used to prove our first "theorem."

Theorem 8.3. *The Yoneda embedding* $y : \mathbf{C} \to \mathbf{Sets}^{\mathbf{C}^{\mathrm{op}}}$ *is full and faithful.*

Proof. For any objects $C, D \in \mathbf{C}$, we have an isomorphism

$$
\mathrm{Hom}_{\mathbf{C}}(C, D) = yD(C) \cong \mathrm{Hom}_{\mathbf{Sets}^{\mathbf{C}^{\mathrm{op}}}}(yC, yD).
$$

And this isomorphism is indeed induced by the functor y, since by (8.3) it takes an element $h : C \to D$ of $yD(C)$ to the natural transformation $\vartheta_h : yC \to yD$ given by

$$
\begin{aligned}
(\vartheta_h)_{C'}(f : C' \to C) &= yD(f)(h) \\
&= \mathrm{Hom}_{\mathbf{C}}(f, D)(h) \\
&= h \circ f \\
&= (yh)_{C'}(f),
\end{aligned}
$$

where $yh : yC \to yD$ has component at C':

$$
\begin{aligned}
(yh)_{C'} : \mathrm{Hom}(C', C) &\longrightarrow \mathrm{Hom}(C', D) \\
f &\longmapsto h \circ f
\end{aligned}
$$

So, $\vartheta_h = y(h)$. $\qquad\qquad\qquad\qquad\qquad\qquad\qquad\qquad\qquad\qquad\square$

Remark 8.4. Note the following:

- If $\mathbf{C}$ is small, then $\mathbf{Sets}^{\mathbf{C}^{\mathrm{op}}}$ is locally small, and so $\mathrm{Hom}(yC, P)$ in $\mathbf{Sets}^{\mathbf{C}^{\mathrm{op}}}$ is a set.

- If $\mathbf{C}$ is locally small, then $\mathbf{Sets}^{\mathbf{C}^{\mathrm{op}}}$ need not be locally small. In this case, the Yoneda lemma tells us that $\mathrm{Hom}(yC, P)$ is always a set.

- If $\mathbf{C}$ is not locally small, then $y : \mathbf{C} \to \mathbf{Sets}^{\mathbf{C}^{\mathrm{op}}}$ will not even be defined, so the Yoneda lemma does not apply.

Finally, observe that the Yoneda embedding $y : \mathbf{C} \to \mathbf{Sets}^{\mathbf{C}^{\mathrm{op}}}$ is also injective on objects. For, given objects A, B in $\mathbf{C}$, if $yA = yB$ then $1_C \in \mathrm{Hom}(C, C) = yC(C) = yD(C) = \mathrm{Hom}(C, D)$ implies $C = D$.

8.4 Applications of the Yoneda lemma

One frequent sort of application of the Yoneda lemma is of the following form: given objects A, B in a category $\mathbf{C}$, to show that $A \cong B$ it suffices to show that $yA \cong yB$ in $\mathbf{Sets}^{\mathbf{C}^{\mathrm{op}}}$. This "Yoneda principle" results from the foregoing theorem and the fact that, if $F : \mathbf{C} \to \mathbf{D}$ is any full and faithful functor, then $FA \cong FB$ clearly implies $A \cong B$. We record this as the following.

Corollary 8.5 (Yoneda principle). *Given objects A and B in any locally small category $\mathbf{C}$,*

$$yA \cong yB \qquad implies \qquad A \cong B.$$

A typical such case is this. In any cartesian closed category $\mathbf{C}$, we know there is always an isomorphism,

$$(A^B)^C \cong A^{(B \times C)},$$

for any objects A, B, C. But recall how involved it was to prove this directly, using the compound universal mapping property (or a lengthy calculation in λ-calculus). Now, however, by the Yoneda principle, we just need to show that

$$y((A^B)^C) \cong y(A^{(B \times C)}).$$

To that end, take any object $X \in \mathbf{C}$; then we have isomorphisms:

$$\mathrm{Hom}(X, (A^B)^C) \cong \mathrm{Hom}(X \times C, A^B)$$
$$\cong \mathrm{Hom}((X \times C) \times B, A)$$
$$\cong \mathrm{Hom}(X \times (B \times C), A)$$
$$\cong \mathrm{Hom}(X, A^{(B \times C)}).$$

Of course, it must be checked that these isomorphisms are natural in X, but that is straightforward. For instance, for the first one suppose we have $f : X' \to X$. Then, the naturality of the first isomorphism means that for any $g : X \to (A^B)^C$, we have

$$\overline{g \circ f} = \bar{g} \circ (f \times 1),$$

which is clearly true by the uniqueness of transposition (the reader should draw the diagram).

Here is another sample application of the Yoneda principle.

Proposition 8.6. *If the cartesian closed category $\mathbf{C}$ has coproducts, then $\mathbf{C}$ is "distributive," that is, there is always a canonical isomorphism,*

$$(A \times B) + (A \times C) \cong A \times (B + C).$$

Proof. As in the previous proposition, we check that

$$\mathrm{Hom}(A \times (B + C), X) \cong \mathrm{Hom}(B + C, X^A)$$
$$\cong \mathrm{Hom}(B, X^A) \times \mathrm{Hom}(C, X^A)$$
$$\cong \mathrm{Hom}(A \times B, X) \times \mathrm{Hom}(A \times C, X)$$
$$\cong \mathrm{Hom}((A \times B) + (A \times C), X).$$

Finally, as in the foregoing example, one sees easily that these isos are all natural in X.   □

We have already used a simple logical version of the Yoneda lemma several times: to show that in the propositional calculus one has $\varphi \dashv\vdash \psi$ for some formulas φ, ψ, it suffices to show that for any formula ϑ, one has $\vartheta \vdash \varphi$ iff $\vartheta \vdash \psi$.

More generally, given any objects A, B in a locally small category $\mathbf{C}$, to find an arrow $h : A \to B$ it suffices to give one $\vartheta : yA \to yB$ in $\mathbf{Sets}^{\mathbf{C}^{\mathrm{op}}}$, for then there is a unique h with $\vartheta = yh$. Why should it be easier to give an arrow $yA \to yB$ than one $A \to B$? The key difference is that in general $\mathbf{Sets}^{\mathbf{C}^{\mathrm{op}}}$ has much more structure to work with than does $\mathbf{C}$; as we see, it is complete, cocomplete, cartesian closed, and more. So one can use various "higher-order" tools, from limits to λ-calculus; and if the result is an arrow of the form $yA \to yB$, then it comes from a unique one $A \to B$, despite the fact that $\mathbf{C}$ itself may not admit the "higher-order" constructions. In that sense, the category $\mathbf{Sets}^{\mathbf{C}^{\mathrm{op}}}$ is like an extension of $\mathbf{C}$ by "ideal elements" that permit calculations which cannot be done in $\mathbf{C}$. This is something like passing to the complex numbers to solve equations in the reals, or adding higher types to an elementary logical theory.

8.5 Limits in categories of diagrams

Recall that a category $\mathcal{E}$ is said to be *complete* if it has all small limits; that is, for any small category J and functor $F : J \to \mathcal{E}$, there is a limit $L = \varprojlim_{j \in J} Fj$ in $\mathcal{E}$ and a "cone" $\eta : \Delta L \to F$ in $\mathcal{E}^J$, universal among arrows from constant functors ΔE. Here, the constant functor $\Delta : \mathcal{E} \to \mathcal{E}^J$ is the transposed projection $\mathcal{E} \times J \to \mathcal{E}$.

Proposition 8.7. *For any locally small category* $\mathbf{C}$, *the functor category* $\mathbf{Sets}^{\mathbf{C}^{\mathrm{op}}}$ *is complete. Moreover, for every object* $C \in \mathbf{C}$, *the evaluation functor*

$$\mathrm{ev}_C : \mathbf{Sets}^{\mathbf{C}^{\mathrm{op}}} \to \mathbf{Sets}$$

preserves all limits.

Proof. Suppose we have J small and $F : J \to \mathbf{Sets}^{\mathbf{C}^{\mathrm{op}}}$. The limit of F, if it exists, is an object in $\mathbf{Sets}^{\mathbf{C}^{\mathrm{op}}}$, hence is a functor,

$$(\varprojlim_{j \in J} F_j) : \mathbf{C}^{\mathrm{op}} \to \mathbf{Sets}.$$

By the Yoneda lemma, if we had such a functor, then for each object $C \in \mathbf{C}$ we would have a natural isomorphism,

$$(\varprojlim F_j)(C) \cong \mathrm{Hom}(yC, \varprojlim F_j).$$

But then it would be the case that

$$\mathrm{Hom}(yC, \varprojlim F_j) \cong \varprojlim \mathrm{Hom}(yC, F_j) \qquad \text{in } \mathbf{Sets}$$

$$\cong \varprojlim F_j(C) \qquad \text{in } \mathbf{Sets}$$

where the first isomorphism is because representable functors preserve limits, and the second is Yoneda again. Thus, we are led to *define* the limit $\varprojlim_{j \in J} F_j$ to be

$$(\varprojlim_{j \in J} F_j)(C) = \varprojlim_{j \in J} (F_j C) \qquad (8.4)$$

that is, the *pointwise limit* of the functors F_j. The reader can easily work out how $\varprojlim F_j$ acts on $\mathbf{C}$-arrows, and what the universal cone is, and our hypothetical argument then shows that it is indeed a limit in $\mathbf{Sets}^{\mathbf{C}^{\mathrm{op}}}$.

Finally, the preservation of limits by evaluation functors is stated by (8.4). $\square$

8.6 Colimits in categories of diagrams

The notion of *cocompleteness* is of course the dual of completeness: a category is cocomplete if it has all (small) colimits. Like the foregoing proposition about the completeness of $\mathbf{Sets}^{\mathbf{C}^{\mathrm{op}}}$, its cocompleteness actually follows simply from the fact that $\mathbf{Sets}$ is cocomplete. We leave the proof of the following as an exercise.

Proposition 8.8. *Given any categories $\mathbf{C}$ and $\mathbf{D}$, if $\mathbf{D}$ is cocomplete, then so is the functor category $\mathbf{D}^{\mathbf{C}}$, and the colimits in $\mathbf{D}^{\mathbf{C}}$ are "computed pointwise," in the sense that for every $C \in \mathbf{C}$, the evaluation functor*

$$\mathrm{ev}_C : \mathbf{D}^{\mathbf{C}} \to \mathbf{D}$$

preserves colimits. Thus, for any small index category J and functor $A : J \to \mathbf{D}^{\mathbf{C}}$, for each $C \in \mathbf{C}$ there is a canonical isomorphism,

$$(\varinjlim_{j \in J} A_j)(C) \cong \varinjlim_{j \in J} (A_j C).$$

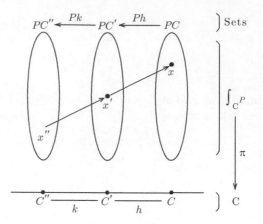

Figure 8.1 Category of elements

Proof. Exercise.   □

Corollary 8.9. *For any locally small* **C***, the functor category* **Sets**$^{C^{op}}$ *is cocomplete, and colimits there are computed pointwise.*

Proposition 8.10. *For any small category* **C***, every object* P *in the functor category* **Sets**$^{C^{op}}$ *is a colimit of representable functors,*

$$\varinjlim_{j \in J} yC_j \cong P.$$

More precisely, there is a canonical choice of an index category J *and a functor* $\pi : J \to$ **C** *such that there is a natural isomorphism* $\varinjlim_J y \circ \pi \cong P$.

Proof. Given $P :$ **C**$^{op} \to$ **Sets**, the index category we need is the so-called *category of elements* of P, written,

$$\int_{\mathbf{C}} P$$

and defined as follows.

Objects: pairs (x, C) where $C \in$ **C** and $x \in PC$.

Arrows: an $h : (x', C') \to (x, C)$ is an arrow $h : C' \to C$ in **C** such that

$$P(h)(x) = x' \tag{8.5}$$

actually, the arrows are triples of the form $(h, (x', C'), (x, C))$ satisfying (8.5).

The reader can easily work out the obvious identities and composites. See Figure 8.1.

Note that $\int_{\mathbf{C}} P$ is a small category since $\mathbf{C}$ is small. There is a "projection" functor,

$$\pi : \int_{\mathbf{C}} P \to \mathbf{C}$$

defined by $\pi(x, C) = C$ and $\pi(h, (x', C'), (x, C)) = h$.

To define the cocone of the form $y \circ \pi \to P$, take an object $(x, C) \in \int_{\mathbf{C}} P$ and observe that (by the Yoneda lemma) there is a natural, bijective correspondence between

$$\frac{x \in P(C)}{x : yC \to P}$$

which we simply identify notationally. Moreover, given any arrow $h : (x', C') \to (x, C)$ naturality in C implies that there is a commutative triangle

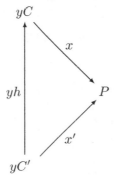

Indeed, the category $\int_{\mathbf{C}} P$ is thus equivalent to the full subcategory of the slice category over P on the objects $yC \to P$ (i.e., arrows in $\mathbf{Sets}^{\mathbf{C}^{\mathrm{op}}}$) with representable domains.

We can therefore take the component of the desired cocone $y\pi \to P$ at (x, C) to be simply $x : yC \to P$. To see that this is a colimiting cocone, take any cocone $y\pi \to Q$ with components $\vartheta_{(x,C)} : yC \to Q$ and we require a unique natural transformation $\vartheta : P \to Q$ as indicated in the following diagram:

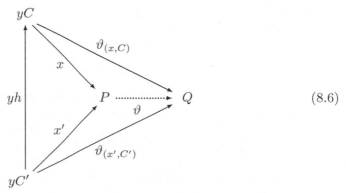

(8.6)

We can define $\vartheta_C : PC \to QC$ by setting

$$\vartheta_C(x) = \vartheta_{(x,C)}$$

where we again identify,

$$\frac{\vartheta_{(x,C)} \in Q(C)}{\vartheta_{(x,C)} : yC \to Q}$$

This assignment is clearly natural in C by the commutativity of the diagram (8.6). For uniqueness, given any $\varphi : P \to Q$ such that $\varphi \circ x = x'$, again by Yoneda we must have $\varphi \circ x = \vartheta_{(x,c)} = \vartheta \circ x$. $\qquad\square$

We include the following because it fits naturally here, but defer the proof to Chapter 9, where a neat proof can be given using adjoint functors. As an exercise, the reader may wish to prove it at this point using the materials already at hand, which is also quite doable.

Proposition 8.11. *For any small category* **C**, *the Yoneda embedding*

$$y : \mathbf{C} \to \mathbf{Sets}^{\mathbf{C}^{\mathrm{op}}}$$

is the "free cocompletion" of **C**, *in the following sense. Given any cocomplete category* $\mathcal{E}$ *and functor* $F : \mathbf{C} \to \mathcal{E}$, *there is a colimit preserving functor* $F_! :$ $\mathbf{Sets}^{\mathbf{C}^{\mathrm{op}}} \to \mathcal{E}$, *unique up to natural isomorphism with the property*

$$F_! \circ y \cong A$$

as indicated in the following diagram:

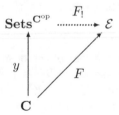

Proof. (Sketch, see proposition 9.16.) Given $F : \mathbf{C} \to \mathcal{E}$, define $F_!$ as follows. For any $P \in \mathbf{Sets}^{\mathbf{C}^{\mathrm{op}}}$, let

$$\varinjlim_{j \in J} yA_j \cong P$$

be the canonical presentation of P as a colimit of representables with $J = \int_{\mathbf{C}} P$, the category of elements of P. Then set,

$$F_!(P) = \varinjlim_{j \in J} F(A_j)$$

which exists since $\mathcal{E}$ is cocomplete. $\qquad\square$

8.7 Exponentials in categories of diagrams

As an application, let us consider exponentials in categories of the form $\mathbf{Sets}^{\mathbf{C}^{\mathrm{op}}}$ for small $\mathbf{C}$. We need the following lemma.

Lemma 8.12. *For any small index category J, functor $A : J \to \mathbf{Sets}^{\mathbf{C}^{\mathrm{op}}}$ and diagram $B \in \mathbf{Sets}^{\mathbf{C}^{\mathrm{op}}}$, there is a natural isomorphism*

$$\varinjlim_{j} (A_j \times B) \cong (\varinjlim_{j} A_j) \times B. \tag{8.7}$$

Briefly, the functor $- \times B : \mathbf{Sets}^{\mathbf{C}^{\mathrm{op}}} \to \mathbf{Sets}^{\mathbf{C}^{\mathrm{op}}}$ preserves colimits.

Proof. To specify the canonical natural transformation mentioned in (8.7), start with the cocone,

$$\vartheta_j : A_j \to \varinjlim_{j} A_j, \qquad j \in J$$

apply the functor $- \times B$ to get a cocone,

$$\vartheta_j \times B : A_j \times B \to (\varinjlim_{j} A_j) \times B, \qquad j \in J$$

and so there is a unique "comparison arrow" from the colimit,

$$\vartheta : \varinjlim_{j}(A_j \times B) \to (\varinjlim_{j} A_j) \times B,$$

which we claim is a natural isomorphism.

By exercise 7 of Chapter 7, it suffices to show that each component,

$$\vartheta_C : (\varinjlim_{j}(A_j \times B))(C) \to ((\varinjlim_{j} A_j) \times B)(C)$$

is iso. But since the limits and colimits involved are all computed pointwise, it therefore suffices to show (8.7) under the assumption that the A_j and B are just sets. To that end, take any set X and consider the following isomorphisms in **Sets**,

$$\mathrm{Hom}(\varinjlim_{j}(A_j \times B), X) \cong \varprojlim_{j} \mathrm{Hom}(A_j \times B, X)$$

$$\cong \varprojlim_{j} \mathrm{Hom}(A_j, X^B) \qquad (\mathbf{Sets} \text{ is CCC})$$

$$\cong \mathrm{Hom}(\varinjlim_{j} A_j, X^B)$$

$$\cong \mathrm{Hom}((\varinjlim_{j} A_j) \times B, X).$$

Since these are natural in X, the claim follows by Yoneda. $\qquad\square$

Now suppose we have functors P and Q and we want Q^P. The reader should try to construct the exponential "pointwise,"

$$Q^P(C) \overset{?}{=} Q(C)^{P(C)}$$

to see that it *does not* work (it is not functorial in C, as the exponent is contravariant in C).

Let us instead reason as follows: if we had such an exponential Q^P, we could compute its value at any object $C \in \mathbf{C}$ by Yoneda:

$$Q^P(C) \cong \mathrm{Hom}(yC, Q^P)$$

And if it is to be an exponential, then we must also have

$$\mathrm{Hom}(yC, Q^P) \cong \mathrm{Hom}(yC \times P, Q).$$

But this latter set *does* exist, and it *is* functorial in C. Thus, we are led to *define*

$$Q^P(C) = \mathrm{Hom}(yC \times P, Q) \qquad (8.8)$$

with the action on $h : C' \to C$ being

$$Q^P(h) = \mathrm{Hom}(yh \times 1_P, Q).$$

This is clearly a contravariant, set-valued functor on $\mathbf{C}$. Let us now check that it indeed gives an exponential of P and Q.

Proposition 8.13. *For any objects X, P, Q in $\mathbf{Sets}^{\mathbf{C}^{\mathrm{op}}}$, there is an isomorphism, natural in X,*

$$\mathrm{Hom}(X, Q^P) \cong \mathrm{Hom}(X \times P, Q).$$

Proof. By proposition 8.10, for a suitable index category J, we can write X as a colimit of representables,

$$X \cong \varinjlim_{j \subset J} yC_j.$$

Thus we have isomorphisms,

$$\mathrm{Hom}(X, Q^P) \cong \mathrm{Hom}(\varinjlim_{j} yC_j, Q^P)$$

$$\cong \varprojlim_{j} \mathrm{Hom}(yC_j, Q^P)$$

$$\cong \varprojlim_{j} Q^P(C_j) \qquad \text{(by Yoneda)}$$

$$\cong \varprojlim_{j} \mathrm{Hom}(yC_j \times P, Q) \qquad \text{(by 8.8)}$$

$$\cong \mathrm{Hom}(\varinjlim_{j}(yC_j \times P), Q)$$

$$\cong \mathrm{Hom}(\varinjlim_{j}(yC_j) \times P, Q) \qquad \text{(Lemma 8.12)}$$

$$\cong \mathrm{Hom}(X \times P, Q).$$

And as usual these isos are clearly natural in X.   □

Theorem 8.14. *For any small category* **C**, *the category of diagrams* **Sets**$^{\mathbf{C}^{\mathrm{op}}}$ *is cartesian closed. Moreover, the Yoneda embedding*

$$y : \mathbf{C} \to \mathbf{Sets}^{\mathbf{C}^{\mathrm{op}}}$$

preserves all products and exponentials that exist in **C**.

Proof. In light of the foregoing proposition, it only remains to show that y preserves products and exponentials. We leave this as an easy exercise.   □

Remark 8.15. As a corollary, we find that we can sharpen the CCC completeness theorem 6.17 for the simply-typed λ-calculus by restricting to CCCs of the special form **Sets**$^{\mathbf{C}^{\mathrm{op}}}$.

8.8 Topoi

Since we are now so close to it, we might as well introduce the important notion of a "topos"—even though this is not the place to develop that theory, as appealing as it is. First we require the following generalization of characteristic functions of subsets.

Definition 8.16. Let $\mathcal{E}$ be a category with all finite limits. A *subobject classifier* in $\mathcal{E}$ consists of an object Ω together with an arrow $t : 1 \to \Omega$ that is a "universal subobject," in the following sense:

Given any object E and any subobject $U \rightarrowtail E$, there is a unique arrow $u : E \to \Omega$ making the following diagram a pullback:

$$(8.9)$$

The arrow u is called the *classifying* arrow of the subobject $U \rightarrowtail E$; it can be thought of as taking exactly the part of E that is U to the "point" t of Ω. The most familiar example of a subobject classifier is of course the set $2 = \{0, 1\}$ with a selected element as $t : 1 \to 2$. The fact that every subset $U \subseteq S$ of any set S has a unique characteristic function $u : S \to 2$ is then exactly the subobject classifier condition.

It is easy to show that a subobject classifier is unique up to isomorphism: the pullback condition is clearly equivalent to requiring the contravariant subobject functor,

$$\mathrm{Sub}_{\mathcal{E}}(-) : \mathcal{E}^{\mathrm{op}} \to \mathbf{Sets}$$

(which acts by pullback) to be representable,

$$\mathrm{Sub}_{\mathcal{E}}(-) \;\cong\; \mathrm{Hom}_{\mathcal{E}}(-, \Omega).$$

The required isomorphism is just the pullback condition stated in the definition of a subobject classifier. Now apply the Yoneda principle, corollary 8.5, for two subobject classifiers Ω and Ω'.

Definition 8.17. A *topos* is a category $\mathcal{E}$ such that

1. $\mathcal{E}$ has all finite limits,
2. $\mathcal{E}$ has a subobject classifier,
3. $\mathcal{E}$ has all exponentials.

This compact definition proves to be amazingly rich in consequences: it can be shown for instance that topoi also have all finite colimits, and that every slice category of a topos is again a topos. We refer the reader to the books by Mac Lane and Moerdijk (1992), Johnstone (2002), and McLarty (1995) for information on topoi, and here just give an example (albeit one that covers a very large number of cases).

Proposition 8.18. *For any small category* $\mathbf{C}$, *the category of diagrams* $\mathbf{Sets}^{\mathbf{C}^{\mathrm{op}}}$ *is a topos.*

Proof. Since we already know that $\mathbf{Sets}^{\mathbf{C}^{\mathrm{op}}}$ has all limits, and we know that it has exponentials by Section 8.7, we just need to find a subobject classifier. To that end, we define a *sieve* on an object C of $\mathbf{C}$ to be a set S of arrows $f : \cdot \to C$ (with arbitrary domain) that is closed under precomposition; that is, if $f : D \to C$ is in S then so is $f \circ g : E \to D \to C$ for every $g : E \to D$ (think of a sieve as a common generalization of a "lower set" in a poset and an "ideal" in a ring). Then let

$$\Omega(C) \;=\; \{S \subseteq \mathbf{C}_1 \mid S \text{ is a sieve on } C\}$$

and given $h : D \to C$, let

$$h^* : \Omega(C) \to \Omega(D)$$

be defined by

$$h^*(S) = \{g : \cdot \to D \mid h \circ g \in S\}.$$

This clearly defines a presheaf $\Omega : \mathbf{C}^{\mathrm{op}} \to \mathbf{Sets}$, with a distinguished point,

$$t : 1 \to \Omega$$

namely, at each C, the "total sieve"

$$t_C = \{f : \cdot \to C\}.$$

We claim that $t : 1 \to \Omega$ so defined is a subobject classifier for $\mathbf{Sets}^{\mathbf{C}^{\mathrm{op}}}$. Indeed, given any object E and subobject $U \rightarrowtail E$, define $u : E \to \Omega$ at any object $C \in \mathbf{C}$ by

$$u_C(e) = \{f : D \to C \mid f^*(e) \in U(D) \rightarrowtail E(D)\}$$

for any $e \in E(C)$. That is, $u_C(e)$ is the sieve of arrows into C that take $e \in E(C)$ back into the subobject U.

$\square$

The notion of a topos first arose in the Grothendieck school of algebraic geometry as a generalization of that of a topological space. But one of the most fascinating aspects of topoi is their relation to logic. In virtue of the association of subobjects $U \rightarrowtail E$ with arrows $u : E \to \Omega$, the subobject classifier Ω can be regarded as an object of "propositions" or "truth-values," with $t = \mathrm{true}$. An arrow $\varphi : E \to \Omega$ is then a "propositional function" of which $U_\varphi \rightarrowtail E$ is the "extension." For, by the pullback condition (8.9), a generalized element $x : X \to E$ is "in" U_φ (i.e., factors through $U_\varphi \rightarrowtail E$) just if $\varphi x = \mathrm{true}$,

$$x \in_E U_\varphi \qquad \text{iff} \qquad \varphi x = \mathrm{true}$$

so that, again in the notation of Section 5.1,

$$U_\varphi = \{x \in E \mid \varphi x = \mathrm{true}\}.$$

This permits an interpretation of first-order logic in any topos, since topoi also have a way of modeling the logical quantifiers $\exists$ and $\forall$ as adjoints to pullbacks (as described in Section 9.5).

Since topoi are also cartesian closed, they have an internal type theory described by the λ-calculus (see Section 6.6). Combining this with the first-order logic and subobject classifier Ω provides a natural interpretation of *higher-order* logic, employing the exponential Ω^E as a "power object" $P(E)$ of subobjects of E. This logical aspect of topoi is also treated in the books already mentioned.

8.9 Exercises

1. If $F : \mathbf{C} \to \mathbf{D}$ is full and faithful, then $C \cong C'$ iff $FC \cong FC'$.

2. Let $\mathbf{C}$ be a small category. Prove that the representable functors *generate* the diagram category $\mathbf{Sets}^{\mathbf{C}^{\mathrm{op}}}$, in the following sense: given any objects $P, Q \in \mathbf{Sets}^{\mathbf{C}^{\mathrm{op}}}$ and natural transformations $\varphi, \psi : P \to Q$, if for every representable functor yC and natural transformation $\vartheta : yC \to P$, one has $\varphi \circ \vartheta = \psi \circ \vartheta$, then $\varphi = \psi$. Thus, the arrows in $\mathbf{Sets}^{\mathbf{C}^{\mathrm{op}}}$ are determined by their effect on generalized elements based at representables.

3. Let $\mathbf{C}$ be a locally small, cartesian closed category. Use the Yoneda embedding to show that for any objects A, B, C in $\mathbf{C}$

$$(A \times B)^C \cong A^C \times B^C$$

(cf. problem 2 Chapter 6).
If **C** also has binary coproducts, show that also

$$A^{(B+C)} \cong A^B \times A^C.$$

4. Let Δ be the category of finite ordinal numbers $0, 1, 2, \ldots$ and order-preserving maps, and write $[-] : \Delta \to \mathbf{Pos}$ for the evident inclusion. For each poset P, define the simplicial set $S(\mathrm{P})$ by

$$S(\mathrm{P})(n) = \mathrm{Hom}_{\mathbf{Pos}}([n], \mathrm{P}).$$

Show that this specification determines a functor $S : \mathbf{Pos} \to \mathbf{Sets}^{\Delta^{\mathrm{op}}}$ into simplicial sets, and that it coincides with the "simplicial nerve" of P as specified in the text. Is S faithful? Show that S preserves all limits.

5. Generalize the foregoing exercise from posets to (locally small) categories to define the simplicial nerve of a category **C**.

6. Let **C** be any category and **D** any complete category. Show that the functor category $\mathbf{D}^{\mathbf{C}}$ is also complete.
Use duality to show that the same is true for cocompleteness in place of completeness.

7. Let **C** be a locally small category with binary products, and show that the Yoneda embedding

$$y : \mathbf{C} \to \mathbf{Sets}^{\mathbf{C}^{\mathrm{op}}}$$

preserves them. (Hint: this involves only a few lines of calculation.)
If **C** also has exponentials, show that y also preserves them.

8. Show that if P is a poset and $A : \mathrm{P}^{\mathrm{op}} \to \mathbf{Sets}$ a presheaf on P, then the category of elements $\int_{\mathrm{P}} A$ is also a poset and the projection $\pi : \int_{\mathrm{P}} A \to \mathrm{P}$ is a monotone map.
Show, moreover, that the assignment $A \mapsto (\pi : \int_{\mathrm{P}} A \to \mathrm{P})$ determines a functor,

$$\int_{\mathrm{P}} : \mathbf{Sets}^{\mathrm{P}^{\mathrm{op}}} \longrightarrow \mathbf{Pos}/\mathrm{P}.$$

9. Let $\mathbb{T}$ be a theory in the λ-calculus. For any type symbols σ and τ, let

$$[\sigma \to \tau] = \{M : \sigma \to \tau \mid M \text{ closed}\}$$

be the set of closed terms of type $\sigma \to \tau$. Suppose that for each type symbol ρ, there is a function,

$$f_\rho : [\rho \to \sigma] \to [\rho \to \tau]$$

with the following properties:

- for any closed terms $M, N : \rho \to \sigma$, if $\mathbb{T} \vdash M = N$ (provable equivalence from $\mathbb{T}$), then $f_\rho M = f_\rho N$,

- for any closed terms $M : \mu \to \nu$ and $N : \nu \to \sigma$,

$$\mathbb{T} \vdash f_\mu(\lambda x : \mu.N(Mx)) = \lambda x : \mu.(f_\nu(N))(Mx)$$

Use the Yoneda embedding of the cartesian closed *category of types* $\mathbf{C}_\mathbb{T}$ of $\mathbb{T}$ to show that there is a term $F : \sigma \to \tau$ such that f_ρ is induced by composition with F, in the sense that, for every closed term $R : \rho \to \sigma$,

$$\mathbb{T} \vdash f_\rho(R) = \lambda x : \rho.F(Rx)$$

Show that, moreover, F is unique up to $\mathbb{T}$-provable equivalence.

10. Combine proposition 6.17 with theorem 8.14 to infer that the λ-calculus is deductively complete with respect to categories of diagrams.

11. Show that every slice category $\mathbf{Sets}/X$ is cartesian closed. Calculate the exponential of two objects $A \to X$ and $B \to X$ by first determining the Yoneda embedding $y : X \to \mathbf{Sets}^X$, and then applying the formula for exponentials of presheaves. Finally, observe that $\mathbf{Sets}/X$ is a topos, and determine its subobject classifier.

12. (a) Explicitly determine the subobject classifiers for the topoi $\mathbf{Sets}^2$ and $\mathbf{Sets}^\omega$, where as always $\mathbf{2}$ is the poset $0 < 1$ and ω is the poset of natural numbers $0 < 1 < 2 < \cdots$.

    (b) Show that $(\mathbf{Sets}_{\text{fin}})^2$ is a topos.

13. Explicitly determine the graph that is the subobject classifier in the topos of graphs (i.e., what are its edges and vertices?). How many points $1 \to \Omega$ does it have?

9

ADJOINTS

This chapter represents the high point of this book, the goal toward which we have been working steadily. The notion of adjoint functor, first discovered by D. Kan in the 1950s, applies everything that we have learned up to now to unify and subsume all of the different universal mapping properties that we have encountered, from free groups to limits to exponentials. But more importantly, it also captures an important mathematical phenomenon that is invisible without the lens of category theory. Indeed, I will make the admittedly provocative claim that adjointness is a concept of fundamental logical and mathematical importance that is not captured elsewhere in mathematics.

Many of the most striking applications of category theory involve adjoints, and many important and fundamental mathematical notions are instances of adjoint functors. As such, they share the common behavior and formal properties of all adjoints, and in many cases this fact alone accounts for all of their essential features.

9.1 Preliminary definition

We begin by recalling the universal mapping property (UMP) of free monoids: every monoid M has an underlying set $U(M)$, and every set X has a free monoid $F(X)$, and there is a function

$$i_X : X \to UF(X)$$

with the following UMP:

For every monoid M and every function $f : X \to U(M)$, there is a unique homomorphism $g : F(X) \to M$ such that $f = U(g) \circ i_X$, all as indicated in the following diagram:

Now consider the following map:

$$\phi : \mathrm{Hom}_{\mathbf{Mon}}(F(X), M) \to \mathrm{Hom}_{\mathbf{Sets}}(X, U(M))$$

defined by

$$g \mapsto U(g) \circ i_X.$$

The UMP given above says exactly that ϕ is an isomorphism,

$$\mathrm{Hom}_{\mathbf{Mon}}(F(X), M) \cong \mathrm{Hom}_{\mathbf{Sets}}(X, U(M)). \qquad (9.1)$$

This bijection (9.1) can also be written schematically as a *two-way rule*:

$$\frac{F(X) \longrightarrow M}{X \longrightarrow U(M)}$$

where one gets from an arrow g of the upper form to one $\phi(g)$ of the lower form
by the recipe

$$\phi(g) = U(g) \circ i_X.$$

We pattern our *preliminary* definition of adjunction on this situation. It is
preliminary because it really only gives half of the picture; in Section 9.2 an
equivalent definition emerges as both more convenient and conceptually clearer.

Definition 9.1 (preliminary). An *adjunction* between categories $\mathbf{C}$ and $\mathbf{D}$
consists of functors

$$F : \mathbf{C} \rightleftarrows \mathbf{D} : U$$

and a natural transformation

$$\eta : 1_{\mathbf{C}} \to U \circ F$$

with the property:

(*) For any $C \in \mathbf{C}$, $D \in \mathbf{D}$, and $f : C \to U(D)$, there exists a unique
$g : FC \to D$ such that

$$f = U(g) \circ \eta_C$$

as indicated in

$$F(C) \dashrightarrow^{g} D$$

$$U(F(C)) \xrightarrow{U(g)} U(D)$$

$$\eta_C \nearrow f$$

$$C$$

Terminology and notation:

- F is called the *left adjoint*, U is called the *right adjoint*, and η is called the *unit* of the adjunction.
- One sometimes writes $F \dashv U$ for "F is left and U right adjoint."
- The statement (*) is the UMP of the unit η.

Note that the situation $F \dashv U$ is a generalization of equivalence of categories, in that a pseudo-inverse is an adjoint. In that case, however, it is the relation between categories that one is interested in. Here, one is concerned with the relation between specific functors. That is to say, it is not the relation on categories "there exists an adjunction," but rather "this functor has an adjoint" that we are concerned with.

Suppose now that we have an adjunction,

$$\mathbf{C} \xrightarrow[F]{\overset{U}{\longleftarrow}} \mathbf{D}.$$

Then, as in the example of monoids, take $C \in \mathbf{C}$ and $D \in \mathbf{D}$ and consider the operation

$$\phi : \mathrm{Hom}_{\mathbf{D}}(FC, D) \to \mathrm{Hom}_{\mathbf{C}}(C, UD)$$

given by $\phi(g) = U(g) \circ \eta_C$. Since, by the UMP of η, every $f : C \to UD$ is $\phi(g)$ for a unique g, just as in our example we see that ϕ is an isomorphism

$$\mathrm{Hom}_{\mathbf{D}}(F(C), D) \cong \mathrm{Hom}_{\mathbf{C}}(C, U(D)) \tag{9.2}$$

which, again, can be displayed as the two-way rule:

$$\frac{F(C) \longrightarrow D}{C \longrightarrow U(D)}$$

Example 9.2. Consider the "diagonal" functor,

$$\Delta : \mathbf{C} \to \mathbf{C} \times \mathbf{C}$$

defined on objects by

$$\Delta(C) = (C, C)$$

and on arrows by

$$\Delta(f : C \to C') = (f, f) : (C, C) \to (C', C').$$

What would it mean for this functor to have a right adjoint? We would need a functor $R : \mathbf{C} \times \mathbf{C} \to \mathbf{C}$ such that for all $C \in \mathbf{C}$ and $(X, Y) \in \mathbf{C} \times \mathbf{C}$, there is a bijection:

$$\frac{\Delta C \longrightarrow (X, Y)}{C \longrightarrow R(X, Y)}$$

That is, we would have

$$\text{Hom}_{\mathbf{C}}(C, R(X,Y)) \cong \text{Hom}_{\mathbf{C}\times\mathbf{C}}(\Delta C, (X,Y))$$
$$\cong \text{Hom}_{\mathbf{C}}(C, X) \times \text{Hom}_{\mathbf{C}}(C, Y).$$

We therefore must have $R(X,Y) \cong X \times Y$, suggesting that Δ has as a right adjoint the product functor $\times : \mathbf{C} \times \mathbf{C} \to \mathbf{C}$,

$$\Delta \dashv \times.$$

The unit η would have the form $\eta_C : C \to C \times C$, so we propose the "diagonal arrow" $\eta_C = \langle 1_C, 1_C \rangle$, and we need to check the UMP indicated in the following diagram:

Indeed, given any $f : C \to X \times Y$, we have unique f_1 and f_2 with $f = \langle f_1, f_2 \rangle$, for which, we then have

$$(f_1 \times f_2) \circ \eta_C = \langle f_1 \pi_1, f_2 \pi_2 \rangle \eta_C$$
$$= \langle f_1 \pi_1 \eta_C, f_2 \pi_2 \eta_C \rangle$$
$$= \langle f_1, f_2 \rangle$$
$$= f.$$

Thus in sum, the functor Δ has a right adjoint if and only if $\mathbf{C}$ has binary products.

Example 9.3. For an example of a different sort, consider the category **Pos** of posets and monotone maps and $\mathcal{C}$**Pos** of cocomplete posets and cocontinuous maps. A poset $\mathcal{C}$ is cocomplete just if it has a join $\bigvee_i c_i$ for every family of elements $(c_i)_{i \in I}$ indexed by a set I, and a monotone map $f : \mathcal{C} \to \mathcal{D}$ is cocontinuous if it preserves all such joins, $f(\bigvee_i c_i) = \bigvee_i f(c_i)$. There is an obvious forgetful functor

$$U : \mathcal{C}\mathbf{Pos} \to \mathbf{Pos}.$$

What would a left adjoint $F \dashv U$ be? There would have to be a monotone map $\eta : P \to UF(P)$ with the property: given any cocomplete poset $\mathcal{C}$ and monotone

$f : P \to U(\mathcal{C})$, there exists a unique cocontinuous $\bar{f} : F(P) \to \mathcal{C}$ such that $f = U(\bar{f}) \circ \eta_P$, as indicated in

$$F(P) \cdots\cdots\cdots\overset{\bar{f}}{\longrightarrow} \mathcal{C}$$

$$UF(P) \xrightarrow{U(\bar{f})} U(\mathcal{C})$$

$$\eta \quad\quad f$$

$$P$$

In this precise sense, such a poset $F(P)$ would be a "free cocompletion" of P, and $\eta : P \, \rightarrow \, UF(P)$ a "best approximation" of P by a cocomplete poset.

We leave it to the reader to show that such a "cocompletion" always exists, namely the poset of *lower sets*,

$$\mathrm{Low}(P) = \{U \subseteq P \mid p' \leq p \in U \text{ implies } p' \in U\}.$$

9.2 Hom-set definition

The following proposition shows that the isomorphism (9.2) is in fact natural in both C and D.

Proposition 9.4. *Given categories and functors,*

$$\mathbf{C} \xleftarrow[F]{U} \mathbf{D}$$

the following conditions are equivalent:

1. *F is left adjoint to U; that is, there is a natural transformation*

$$\eta : 1_{\mathbf{C}} \to U \circ F$$

   that has the UMP of the unit:

   For any $C \in \mathbf{C}$, $D \in \mathbf{D}$ and $f : C \to U(D)$, there exists a unique $g : FC \to D$ such that

$$f = U(g) \circ \eta_C.$$

2. *For any $C \in \mathbf{C}$ and $D \in \mathbf{D}$, there is an isomorphism,*

$$\phi : \mathrm{Hom}_{\mathbf{D}}(FC, D) \cong \mathrm{Hom}_{\mathbf{C}}(C, UD)$$

   that is natural in both C and D.

Moreover, the two conditions are related by the formulas

$$\phi(g) = U(g) \circ \eta_C$$
$$\eta_C = \phi(1_{FC}).$$

Proof. (1 implies 2) The recipe for ϕ, given η is just the one stated and we have already observed it to be an isomorphism, given the UMP of the unit. For naturality in C, take $h : C' \to C$ and consider the following diagram:

$$
\begin{array}{ccc}
\mathrm{Hom}_{\mathbf{D}}(FC, D) & \xrightarrow[\cong]{\phi_{C,D}} & \mathrm{Hom}_{\mathbf{C}}(C, UD) \\
{\scriptstyle (Fh)^*} \downarrow & & \downarrow {\scriptstyle h^*} \\
\mathrm{Hom}_{\mathbf{D}}(FC', D) & \xrightarrow[\phi_{C',D}]{\cong} & \mathrm{Hom}_{\mathbf{C}}(C', UD)
\end{array}
$$

Then for any $f : FC \to D$, we have

$$
\begin{aligned}
h^*(\phi_{C,D}(f)) &= (U(f) \circ \eta_C) \circ h \\
&= U(f) \circ UF(h) \circ \eta_{C'} \\
&= U(f \circ F(h)) \circ \eta_{C'} \\
&= \phi_{C',D}(F(h)^*(f)).
\end{aligned}
$$

For naturality in D, take $g : D \to D'$ and consider the diagram

$$
\begin{array}{ccc}
\mathrm{Hom}_{\mathbf{D}}(FC, D) & \xrightarrow[\cong]{\phi_{C,D}} & \mathrm{Hom}_{\mathbf{C}}(C, UD) \\
{\scriptstyle g_*} \downarrow & & \downarrow {\scriptstyle (Ug)_*} \\
\mathrm{Hom}_{\mathbf{D}}(FC, D') & \xrightarrow[\phi_{C,D'}]{\cong} & \mathrm{Hom}_{\mathbf{C}}(C, UD')
\end{array}
$$

Then for any $f : FC \to D$ we have

$$
\begin{aligned}
(Ug)_*(\phi_{C,D}(f)) &= U(g) \circ (U(f) \circ \eta_C) \\
&= U(g \circ f) \circ \eta_C \\
&= \phi_{C,D'}(g \circ f) \\
&= \phi_{C,D'}(g_*(f)).
\end{aligned}
$$

So ϕ is indeed natural.

(2 implies 1) We are given a bijection ϕ,

$$
\frac{F(C) \longrightarrow D}{C \longrightarrow U(D)}
\tag{9.3}
$$

for each C, D, that is natural in C and D. In detail, this means that given a commutative triangle

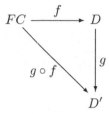

there are two ways to get an arrow of the form $C \to UD'$, namely

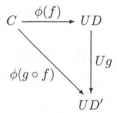

Naturality in D means that this diagram commutes,

$$\phi(g \circ f) = Ug \circ \phi(f). \tag{9.4}$$

Dually, naturality in C means that given

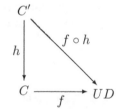

and writing $\psi = \phi^{-1}$, the following commutes:

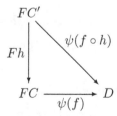

That is,

$$\psi(f \circ h) = \psi(f) \circ Fh.$$

Now, given such a natural bijection ϕ, we want a natural transformation

$$\eta : 1_{\mathbf{C}} \to U \circ F$$

with the UMP of the unit. To find

$$\eta_C : C \to UFC$$

put FC for D and $1_{FC} : FC \to FC$ in the adjoint schema (9.3) to get

$$\frac{1_{FC}: \quad FC \longrightarrow FC}{\eta_C: \quad C \longrightarrow UFC} \; \phi$$

That is, define

$$\eta_C = \phi(1_{FC}).$$

We leave it as an exercise to show that η so defined really is natural in C. Finally, to see that η has the required UMP of the unit, it clearly suffices to show that for all $g : FC \to D$, we have

$$\phi(g) = Ug \circ \eta_C$$

since we are assuming that ϕ is iso. But, using (9.4),

$$Ug \circ \eta_C = Ug \circ \phi(1_{FC})$$
$$= \phi(g \circ 1_{FC})$$
$$= \phi(g).$$

$\square$

Note that the second condition in the foregoing proposition is symmetric, but the first condition is not. This implies that we also have the following dual proposition.

Corollary 9.5. *Given categories and functors*

$$\mathbf{C} \underset{F}{\overset{U}{\longleftarrow\!\!\!-\!\!\!-\!\!\!-}} \mathbf{D}$$

the following conditions are equivalent:

1. *For any $C \in \mathbf{C}$, $D \in \mathbf{D}$, there is an isomorphism*

$$\phi : \mathrm{Hom}_{\mathbf{D}}(FC, D) \cong \mathrm{Hom}_{\mathbf{C}}(C, UD)$$

   that is natural in C and D.

2. *There is a natural transformation*

$$\epsilon : F \circ U \to 1_{\mathbf{D}}$$

   with the following UMP:

   For any $C \in \mathbf{C}$, $D \in \mathbf{D}$ and $g : F(C) \to D$, there exists a unique $f : C \to UD$ such that

$$g = \epsilon_D \circ F(f)$$

as indicated in the following diagram:

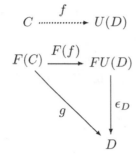

Moreover, the two conditions are related by the equations

$$\psi(f) = \epsilon_D \circ F(f)$$
$$\epsilon_D = \psi(1_{UD})$$

where $\psi = \phi^{-1}$.

Proof. Duality. □

We take the symmetric "Hom-set" formulation as our "official" definition of an adjunction.

Definition 9.6 "official." An *adjunction* consists of functors

$$F : \mathbf{C} \rightleftarrows \mathbf{D} : U$$

and a natural isomorphism

$$\phi : \mathrm{Hom}_{\mathbf{D}}(FC, D) \cong \mathrm{Hom}_{\mathbf{C}}(C, UD) : \psi.$$

This definition has the virtue of being symmetric in F and U. The unit $\eta : 1_{\mathbf{C}} \to U \circ F$ and the *counit* $\epsilon : F \circ U \to 1_{\mathbf{D}}$ of the adjunction are then determined as

$$\eta_C = \phi(1_{FC})$$
$$\epsilon_D = \psi(1_{UD}).$$

9.3 Examples of adjoints

Example 9.7. Suppose $\mathbf{C}$ has binary products. Take a fixed object $A \in \mathbf{C}$, and consider the product functor

$$- \times A : \mathbf{C} \to \mathbf{C}$$

defined on objects by

$$X \mapsto X \times A$$

and on arrows by

$$(h : X \to Y) \mapsto (h \times 1_A : X \times A \longrightarrow Y \times A).$$

When does $- \times A$ have a right adjoint?

We would need a functor

$$U : \mathbf{C} \to \mathbf{C}$$

such that for all $X, Y \in \mathbf{C}$, there is a natural bijection

$$\frac{X \times A \longrightarrow Y}{X \longrightarrow U(Y)}$$

So let us try defining U by

$$U(Y) = Y^A$$

on objects, and on arrows by

$$U(g : Y \to Z) = g^A : Y^A \longrightarrow Z^A.$$

Putting $U(Y)$ for X in the adjunction schema given above then gives the counit:

$$\frac{Y^A \times A \xrightarrow{\ \epsilon\ } Y}{Y^A \xrightarrow[1]{} Y^A}$$

This is, therefore, an adjunction if there is always such a map ϵ with the following UMP:

For any $f : X \times A \to Y$, there is a unique $\bar{f} : X \to Y^A$ such that $f = \epsilon \circ (\bar{f} \times 1_A)$.

But this is exactly the UMP of the exponential! Thus, we do indeed have an adjunction:

$$(-) \times A \dashv (-)^A$$

Example 9.8. Here is a much more simple example. For any category $\mathbf{C}$, consider the unique functor to the terminal category $\mathbf{1}$,

$$! : \mathbf{C} \to \mathbf{1}.$$

Now we ask, when does ! have a right adjoint? This would be an object $U : \mathbf{1} \to \mathbf{C}$ such that for any $C \in \mathbf{C}$, there is a bijective correspondence,

$$\frac{!C \longrightarrow *}{C \longrightarrow U(*)}$$

Such a U would have to be a terminal object in $\mathbf{C}$. So ! has a right adjoint iff $\mathbf{C}$ has a terminal object. What would a left adjoint be?

This last example is a clear case of the following general fact.

Proposition 9.9. *Adjoints are unique up to isomorphism. Specifically, given a functor $F : \mathbf{C} \to \mathbf{D}$ and right adjoints $U, V : \mathbf{D} \to \mathbf{C}$,*

$$F \dashv U \quad and \quad F \dashv V$$

we then have $U \cong V$.

Proof. Here is the easy way. For any $D \in \mathbf{D}$, and $C \in \mathbf{C}$, we have

$$\mathrm{Hom}_{\mathbf{C}}(C, UD) \cong \mathrm{Hom}_{\mathbf{D}}(FC, D) \quad \text{naturally, since } F \dashv U$$

$$\cong \mathrm{Hom}_{\mathbf{C}}(C, VD) \quad \text{naturally, since } F \dashv V.$$

Thus, by Yoneda, $UD \cong VD$. But this isomorphism is natural in D, again by adjointness.   □

This proposition implies that one can use the condition of being right or left adjoint to a given functor to define (uniquely characterize up to isomorphism) a new functor. This sort of characterization, like a UMP, determines an object or construction "structurally" or "intrinsically," in terms of its relation to some other given construction. Many important constructions turn out to be adjoints to particularly simple ones.

For example, what do you suppose would be a *left* adjoint to the diagonal functor

$$\Delta : \mathbf{C} \to \mathbf{C} \times \mathbf{C}$$

in the earlier example 9.2, where $\Delta(C) = (C, C)$ and we had $\Delta \dashv \times$? It would have to be functor $L(X, Y)$ standing in the correspondence

$$\frac{L(X, Y) \longrightarrow C}{(X, Y) \longrightarrow (C, C)}$$

Thus, it could only be the coproduct $L(X, Y) = X + Y$. Therefore, Δ has a left adjoint if and only if $\mathbf{C}$ has binary coproducts,

$$+ \dashv \Delta.$$

Next, note that $\mathbf{C} \times \mathbf{C} \cong \mathbf{C}^2$ where 2 is the discrete two-object category (i.e., any two-element set). Then $\Delta(C)$ is the constant C-valued functor, for each $C \in \mathbf{C}$. Let us now replace 2 by any small index category $\mathbf{J}$ and consider possible adjoints to the corresponding diagonal functor

$$\Delta_{\mathbf{J}} : \mathbf{C} \to \mathbf{C}^{\mathbf{J}}$$

with $\Delta_{\mathbf{J}}(C)(j) = C$ for all $C \in \mathbf{C}$ and $j \in \mathbf{J}$. In this case, one has left and right adjoints

$$\varinjlim_{\mathbf{J}} \ \dashv \ \Delta_{\mathbf{J}} \ \dashv \ \varprojlim_{\mathbf{J}}$$

if and only if **C** has colimits and limits, respectively, of type **J**. Thus, all particular limits and colimits we met earlier, such as pullbacks and coequalizers are
instances of adjoints. What are the units and counits of these adjunctions?

Example 9.10. *Polynomial rings:* Let R be a commutative ring ($\mathbb{Z}$ if you like) and
consider the ring $R[x]$ of polynomials in one indeterminate x with coefficients in
R. The elements of $R[x]$ all look like this:

$$r_0 + r_1 x + r_2 x^2 + \cdots + r_n x^n \qquad (9.5)$$

with the coefficients $r_i \in R$. Of course, there may be some identifications between
such expressions depending on the ring R.

There is an evident homomorphism $\eta : R \to R[x]$, taking elements r to
constant polynomials $r = r_0$, and this map has the following UMP:

> Given any ring A, homomorphism $\alpha : R \to A$, and element $a \in A$, there is
> a unique homomorphism
>
> $$a^* : R[x] \to A$$
>
> such that $a^*(x) = a$ and $a^*\eta = \alpha$.

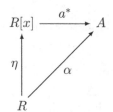

Namely, for a^*, we take the "formal evaluation at a"

$$a^*(r(x)) = \alpha(r)(a/x)$$

given by applying α to the coefficients r_i, substituting a for x, and evaluating
the result in A,

$$a^*(r_0 + r_1 x + r_2 x^2 + \cdots + r_n x^n) = \alpha(r_0) + \alpha(r_1)a + \alpha(r_2)a^2 + \cdots + \alpha(r_n)a^n.$$

To describe this in terms of adjoints, define **Rings**$_*$ to be the category of
"pointed" rings, with objects of the form (A, a), where A is a ring and $a \in A$,
and arrows $h : (A, a) \to (B, b)$ are homomorphisms $h : A \to B$ that preserve the
distinguished point, $h(a) = b$. (Cf. pointed sets, example 7.27.)

The UMP just given says exactly that the functor

$$U : \mathbf{Rings}_* \to \mathbf{Rings}$$

that "forgets the point" $U(A, a) = A$ has as left adjoint the functor

$$[x] : \mathbf{Rings} \to \mathbf{Rings}_*$$

that "adjoins an indeterminate"

$$[x](R) = (R[x], x)$$

and $\eta : R \to R[x]$ is the unit of the adjunction. The reader should have no difficulty working out the details of this example. This provides a characterization of the polynomial ring $R[x]$ by adjointness, one that does not depend on the somewhat vague description in terms of "formal polynomial expressions" like (9.5).

9.4 Order adjoints

Let P be a *preordered set*, that is, a category in which there is at most one arrow $x \to y$ between any two objects. A poset is a preorder that is skeletal. As usual, we define an ordering relation on the objects of P by

$$x \leq y \quad \text{iff there exists an arrow } x \to y.$$

Given another such preorder Q, suppose we have adjoint functors:

$$P \underset{U}{\overset{F}{\rightleftarrows}} Q \qquad\qquad F \dashv U$$

Then the correspondence $Q(Fa, x) \cong P(a, Ux)$ comes down to the simple condition $Fa \leq x$ iff $a \leq Ux$. Thus, an adjunction on preorders consists simply of order-preserving maps F, U satisfying the two-way rule or "bicondition":

$$\frac{Fa \leq x}{a \leq Ux}$$

For each $p \in P$, the unit is therefore an element $p \leq UFp$ that is least among all x with $p \leq Ux$. Dually, for each $q \in Q$ the counit is an element $FUq \leq q$ that is greatest among all y with $Fy \leq q$.

Such a setup on preordered sets is sometimes called a *Galois connection*.

Example 9.11. A basic example is the interior operation on the subsets of a topological space X. Let $\mathcal{O}(X)$ be the set of open subsets of X and consider the operations of inclusion of the opens into the powerset $\mathcal{P}(X)$, and interior:

$$\text{inc} : \mathcal{O}(X) \to \mathcal{P}(X)$$

$$\text{int} : \mathcal{P}(X) \to \mathcal{O}(X)$$

For any subset A and open subset U, the valid bicondition

$$\frac{U \subseteq A}{U \subseteq \text{int}(A)}$$

means that the interior operation is right adjoint to the inclusion of the open subsets among all the subsets:

$$\text{inc} \dashv \text{int}$$

The counit here is the inclusion $\text{int}(A) \subseteq A$, valid for all subsets A. The case of closed subsets and the closure operation is dual.

Example 9.12. A related example is the adjunction on powersets induced by any function $f : A \to B$, between the inverse image operation f^{-1} and the direct image $\text{im}(f)$,

$$\mathcal{P}(A) \xrightarrow[\text{im}(f)]{\overset{f^{-1}}{\longleftarrow}} \mathcal{P}(B)$$

Here we have an adjunction $\text{im}(f) \dashv f^{-1}$ as indicated by the bicondition

$$\frac{\text{im}(f)(U) \subseteq V}{U \subseteq f^{-1}(V)}$$

which is plainly valid for all subsets $U \subseteq A$ and $V \subseteq B$.

The inverse image operation $f^{-1} : \mathcal{P}(B) \to \mathcal{P}(A)$ also has a *right* adjoint, sometimes called the *dual image*, given by

$$f_*(U) = \{b \in B \mid f^{-1}(b) \subseteq U\}$$

which we leave for the reader to verify.

Note that if A and B are topological spaces and $f : A \to B$ is continuous, then f^{-1} restricts to the open sets $f^{-1} : \mathcal{O}(B) \to \mathcal{O}(A)$. Now the left adjoint $\text{im}(f)$ need not exist (on opens), but the right adjoint f_* still does.

$$\mathcal{O}(A) \xrightarrow[f_*]{\overset{f^{-1}}{\longleftarrow}} \mathcal{O}(B)$$

Example 9.13. Suppose we have a poset P. Then, as we know, P has *meets* iff for all $p, q \in P$, there is an element $p \wedge q \in P$ satisfying the bicondition

$$\frac{r \leq p \wedge q}{r \leq p \text{ and } r \leq q}$$

Dually, P has *joins* if there is always an element $p \vee q \in P$ such that

$$\frac{p \vee q \leq r}{p \leq r \text{ and } q \leq r}$$

The *Heyting implication* $q \Rightarrow r$ is characterized as an exponential by the bicondition

$$\frac{p \wedge q \leq r}{p \leq q \Rightarrow r}$$

Finally, an initial object 0 and a terminal object 1 are determined by the conditions

$$0 \leq p$$

and

$$p \leq 1.$$

In this way, the notion of a Heyting algebra can be formulated entirely in terms of adjoints. Equivalently, the intuitionistic propositional calculus is neatly axiomatized by the "adjoint rules of inference" just given (replace "$\leq$" by "$\vdash$"). Together with the reflexivity and transitivity of entailment $p \vdash q$, these rules are completely sufficient for the propositional logical operations. That is, they can serve as the rules of inference for a logical calculus of "binary sequents" $p \vdash q$, which is equivalent to the usual intuitionistic propositional calculus.

When we furthermore define negation by $\neg p = p \Rightarrow \bot$, we then get the derived rule

$$\frac{q \leq \neg p}{p \wedge q \leq 0}$$

Finally, the classical propositional calculus (resp. the laws of Boolean algebra) result from adding the rule

$$\neg\neg p \leq p.$$

Let us now consider how this adjoint analysis of propositional can be extended to all of first-order logic.

9.5 Quantifiers as adjoints

Traditionally, the main obstacle to the further development of *algebraic logic* has been the treatment of the quantifiers. Categorical logic solves this problem beautifully with the recognition (due to F.W. Lawvere in the 1960s) that they, too, are adjoint functors.

Let $\mathcal{L}$ be a first-order language. For any list $\bar{x} = x_1, \ldots, x_n$ of distinct variables let us denote the set of formulas with at most those variables free by

$$\mathrm{Form}(\bar{x}) = \{\phi(\bar{x}) \mid \phi(\bar{x}) \text{ has at most } \bar{x} \text{ free}\}.$$

Then, $\mathrm{Form}(\bar{x})$ is a preorder under the entailment relation of first-order logic

$$\phi(\bar{x}) \vdash \psi(\bar{x}).$$

Now let y be a variable not in the list $\bar{x}$, and note that we have a trivial operation

$$* : \mathrm{Form}(\bar{x}) \rightarrow \mathrm{Form}(\bar{x}, y)$$

taking each $\phi(\bar{x})$ to itself; this is just a matter of observing that if $\phi(\bar{x}) \in \mathrm{Form}(\bar{x})$ then y cannot be free in $\phi(\bar{x})$. Of course, $*$ is trivially a functor, since,

$$\phi(\bar{x}) \vdash \psi(\bar{x}) \quad \text{in } \mathrm{Form}(\bar{x})$$

trivially implies

$$*\phi(\bar{x}) \vdash *\psi(\bar{x}) \quad \text{in Form}(\bar{x}, y).$$

Now since for any $\psi(\bar{x}, y) \in \text{Form}(\bar{x}, y)$ there is, of course, no free y in the formula $\forall y.\psi(\bar{x}, y)$, we have a map

$$\forall y : \text{Form}(\bar{x}, y) \rightarrow \text{Form}(\bar{x}).$$

We claim that this map is right adjoint to $*$,

$$* \dashv \forall.$$

Indeed, the usual rules of universal introduction and elimination imply that the following two-way rule of inference holds:

$$\frac{*\phi(\bar{x}) \vdash \psi(\bar{x}, y)}{\phi(\bar{x}) \vdash \forall y.\psi(\bar{x}, y)} \quad \frac{\text{Form}(\bar{x}, y)}{\text{Form}(\bar{x})}$$

The inference downward is just the usual $\forall$-introduction rule, since y cannot occur freely in $\phi(\bar{x})$. And the inference going up follows from the $\forall$-elimination axiom,

$$\forall y.\psi(\bar{x}, y) \vdash \psi(\bar{x}, y). \tag{9.6}$$

Observe that this derived rule saying that the operation $\forall y$, which binds the variable y, is right adjoint to the trivial operation $*$ depends essentially on the usual "bookkeeping" side condition on the quantifier rule.

Conversely, we could instead take this adjoint rule as basic and derive the customary introduction and elimination rules from it. Indeed, the $\forall$-elimination (9.6) is just the counit of the adjunction, and $\forall$-introduction including the usual side condition results directly from the adjunction.

It is now natural to wonder about the other quantifier $\exists$ of existence; indeed, we have a further adjunction

$$\exists \dashv * \dashv \forall$$

since the following two-way rule also holds:

$$\frac{\exists y.\psi(\bar{x}, y) \vdash \phi(\bar{x})}{\psi(\bar{x}, y) \vdash *\phi(\bar{x})}$$

Here the unit is the existential introduction "axiom"

$$\psi(\bar{x}, y) \vdash \exists y.\psi(\bar{x}, y),$$

and the inference upward is the conventional rule of $\exists$-elimination. It actually follows from these rules that $\exists y$ and $\forall y$ are in particular functors, that is, that $\psi \vdash \phi$ implies $\exists y.\psi \vdash \exists y.\phi$ and similarly for $\forall$.

The adjoint rules just given can thus be used in place of the customary introduction and elimination rules, to give a complete system of deduction for quantificational logic. We emphasize that the somewhat tiresome bookkeeping side conditions typical of the usual logical formulation turn out to be of the essence, since they express the "change of variable context" to which quantifiers are adjoints.

Many typical laws of predicate logic are just simple formal manipulations of adjoints. For example

$$\forall x.\psi(x, y) \vdash \psi(x, y) \qquad \text{(counit of } * \dashv \forall)$$
$$\psi(x, y) \vdash \exists y.\psi(x, y) \qquad \text{(unit of } \exists \dashv *)$$
$$\forall x.\psi(x, y) \vdash \exists y.\psi(x, y) \qquad \text{(transitivity of } \vdash)$$
$$\exists y \forall x.\psi(x, y) \vdash \exists y.\psi(x, y) \qquad (\exists \dashv *)$$
$$\exists y \forall x.\psi(x, y) \vdash \forall x \exists y.\psi(x, y) \qquad (* \dashv \forall)$$

The recognition of the quantifiers as adjoints also gives rise to the following *geometric interpretation*. Take any $\mathcal{L}$ structure M and consider a formula $\phi(x)$ in at most one variable x. It determines a subset,

$$[\phi(x)]^M = \{m \in M \mid M \models \phi(m)\} \subseteq M$$

of all elements satisfying the condition expressed by ϕ. Similarly, a formula in several variables determines a subset of the cartesian product

$$[\psi(x_1, \ldots, x_n)]^M = \{(m_1, \ldots, m_n) \mid M \models \psi(m_1, \ldots, m_n)\} \subseteq M^n.$$

For instance, $[x = y]^M$ is the diagonal subset $\{(m, m) \mid m \in M\} \subseteq M \times M$. Let us take two variables x, y and consider the effect of the $*$ operation on these subsets. The assignment $*[\phi(x)] = [*\phi(x)]$ determines a functor

$$* : \mathcal{P}(M) \to \mathcal{P}(M \times M).$$

Explicitly, given $[\phi(x)] \in \mathcal{P}(M)$, we have

$$*[\phi(x)] = \{(m_1, m_2) \in M \times M \mid M \models \phi(m_1)\} = \pi^{-1}([\phi(x)])$$

where $\pi : M \times M \to M$ is the first projection. Thus,

$$* = \pi^{-1},$$

the inverse image under projection. Similarly, the existential quantifier can be regarded as an operation on subsets by $\exists[\psi(x, y)] = [\exists y.\psi(x, y)]$,

$$\exists : \mathcal{P}(M \times M) \to \mathcal{P}(M).$$

Specifically, given $[\psi(x, y)] \subseteq M \times M$, we have

$$\exists[\psi(x, y)] = [\exists y.\psi(x, y)]$$
$$= \{m \mid \text{ for some } y, M \models \psi(m, y)\}$$
$$= \mathrm{im}(\pi)[\psi(x, y)].$$

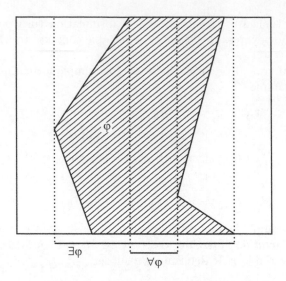

Figure 9.1 Quantifiers as adjoints

Therefore,

$$\exists = \mathrm{im}(\pi),$$

the direct image under projection. In this way, you can actually "see" the logical adjunction:

$$\frac{\exists y.\psi(x,y) \vdash \phi(x)}{\psi(x,y) \vdash \phi(x)}$$

It is essentially the adjunction already considered (example 9.12) between direct and inverse images, applied to the case of a product projection $\pi : M \times M \to M$,

$$\mathrm{im}(\pi) \dashv \pi^{-1}.$$

See Figure 9.1.

Finally, the universal quantifier can also be regarded as an operation of the form

$$\forall : \mathcal{P}(M \times M) \to \mathcal{P}(M)$$

by setting $\forall[\psi(x,y)] = [\forall y.\psi(x,y)]$. Then given $[\psi(x,y)] \subseteq M \times M$, we have

$$\begin{aligned}
\forall[\psi(x,y)] &= [\forall y.\psi(x,y)] \\
&= \{m \mid \text{ for all } y, M \models \psi(m,y)\} \\
&= \{m \mid \pi^{-1}\{m\} \subseteq [\psi(x,y)]\} \\
&= \pi_*([\psi(x,y)]).
\end{aligned}$$

Therefore,

$$\forall = \pi_*$$

so the universal quantifier is the "dual image," that is, the right adjoint to pullback along the projection π. Again, in Figure 9.1, one can see the adjunction:

$$\frac{\phi(x) \leq \psi(x,y)}{\phi(x) \leq \forall y.\psi(x,y)}$$

by considering the corresponding operations induced on subsets.

9.6 RAPL

In addition to the conceptual unification achieved by recognizing constructions as different as existential quantifiers and free groups as instances of adjoints, there is the practical benefit that one then knows that these operations behave in certain ways that are common to all adjoints. We next consider one of the fundamental properties of adjoints: preservation of limits.

In Section 9.5, we had a string of three adjoints,

$$\exists \dashv * \dashv \forall$$

and it is easy to find other such strings. For example, there is a string of four adjoints between **Cat** and **Sets**,

$$V \dashv F \dashv U \dashv R$$

where $U : \textbf{Cat} \rightarrow \textbf{Sets}$ is the forgetful functor to the set of objects

$$U(\textbf{C}) = \textbf{C}_0.$$

An obvious question in this kind of situation is "are there more?" That is, given a functor does it have an adjoint? A useful necessary condition which shows that, for example, the strings above stop is the following proposition, which is also important in its own right.

Proposition 9.14. *Right adjoints preserve limits (RAPL!), and left adjoints preserve colimits.*

Proof. Here is the easy way: suppose we have an adjunction

$$\textbf{C} \underset{U}{\overset{F}{\rightleftarrows}} \textbf{D} \qquad\qquad F \dashv U$$

and we are given a diagram $D : J \rightarrow \textbf{D}$ such that the limit $\varprojlim D_j$ exists in $\textbf{D}$. Then for any $X \in \textbf{C}$, we have

$$\operatorname{Hom}_{\textbf{C}}(X, U(\varprojlim D_j)) \cong \operatorname{Hom}_{\textbf{D}}(FX, \varprojlim D_j)$$

$$\cong \varprojlim \operatorname{Hom}_{\textbf{D}}(FX, D_j)$$

$$\cong \varprojlim \operatorname{Hom}_{\textbf{C}}(X, UD_j)$$

$$\cong \operatorname{Hom}_{\textbf{C}}(X, \varprojlim UD_j)$$

whence (by Yoneda), we have the required isomorphism

$$U(\varprojlim D_j) \cong \varprojlim U D_j.$$

It follows by duality that left adjoints preserve colimits. □

It is illuminating to work out what the above argument "really means" in a particular case, say binary products. Given a product $A \times B$ in $\mathbf{D}$, consider the following diagram, in which the part on the left is in $\mathbf{C}$ and that on the right in $\mathbf{D}$:

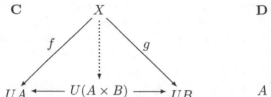 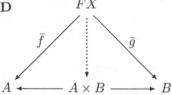

Then given any f and g in $\mathbf{C}$ as indicated, we get the required unique arrow

$\langle f, g \rangle$ by adjointness as the transpose

$$\langle f, g \rangle = \overline{\langle \bar{f}, \bar{g} \rangle}$$

where we write $\bar{f}$, etc., for transposition in both directions.

For an example, recall that in the proof that $\mathbf{Sets}^{\mathbf{C}^{\mathrm{op}}}$ has exponentials we needed the following distributivity law for sets:

$$(\varinjlim_i X_i) \times A \cong \varinjlim_i (X_i \times A)$$

We now see that this is a consequence of the fact that the functor $(-) \times A$ is a left adjoint (namely to $(-)^A$) and therefore preserves colimits.

It also follows immediately for the propositional calculus (and in any Heyting algebra) that, for example,

$$p \Rightarrow (a \wedge b) \dashv\vdash (p \Rightarrow a) \wedge (p \Rightarrow b)$$

and

$$(a \vee b) \wedge p \dashv\vdash (a \wedge p) \vee (b \wedge p).$$

Similarly, for the quantifiers one has, for example,

$$\forall x(\phi(x) \wedge \psi(x)) \dashv\vdash \forall x \phi(x) \wedge \forall x \psi(x).$$

Note that since this does not hold for $\exists x$, it cannot be a right adjoint to some other "quantifier." Similarly

$$\exists x(\phi(x) \vee \psi(x)) \dashv\vdash \exists x \phi(x) \vee \exists x \psi(x).$$

And, as above, $\forall x$ cannot be a left adjoint, since it does not have this property.

The proposition gives an extremely important and useful property of adjoints. As in the foregoing examples, it can be used to show that a given functor does not have an adjoint by showing that it does not preserve (co)limits. But also, to show that a given functor does preserve all (co)limits, sometimes the easiest way to proceed is to show that it has an adjoint. For example, it is very easy to recognize that the forgetful functor $U : \mathbf{Pos} \to \mathbf{Sets}$ from posets to sets has a left adjoint (what is it?). Thus, we know that limits of posets are limits of the underlying sets (suitably ordered). Dually, you may have shown "by hand" as an exercise that the coproduct of free monoids is the free monoid on the coproduct of their generating sets

$$F(A) + F(B) \cong F(A + B).$$

This now follows simply from the free $\dashv$ forgetful adjunction.

Example 9.15. Our final example of preservation of (co)limits by adjoints involves the UMP of the categories of diagrams $\mathbf{Sets}^{\mathbf{C}^{\mathrm{op}}}$ studied in Chapter 8. For a small category $\mathbf{C}$, a contravariant functor $P : \mathbf{C}^{\mathrm{op}} \to \mathbf{Sets}$ is often called a *presheaf* on $\mathbf{C}$, and the functor category $\mathbf{Sets}^{\mathbf{C}^{\mathrm{op}}}$ is accordingly called the *category of presheaves* on $\mathbf{C}$, sometimes written as $\hat{\mathbf{C}}$. This cocomplete category is the "free cocompletion" of $\mathbf{C}$ in the following sense.

Proposition 9.16. *For any small category* $\mathbf{C}$, *the Yoneda embedding*

$$y : \mathbf{C} \to \mathbf{Sets}^{\mathbf{C}^{\mathrm{op}}}$$

has the following UMP: given any cocomplete category $\mathcal{E}$
and functor $F : \mathbf{C} \to \mathcal{E}$, *there is a colimit preserving functor* $F_! : \mathbf{Sets}^{\mathbf{C}^{\mathrm{op}}} \to \mathcal{E}$
such that

$$F_! \circ y \cong F \tag{9.7}$$

as indicated in the following diagram:

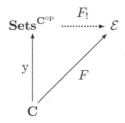

Moreover, up to natural isomorphism, $F_!$ *is the unique cocontinuous functor with this property.*

Proof. We show that there are adjoint functors,

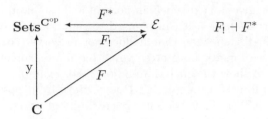

with $F_! \circ y \cong F$. It then follows that $F_!$ preserves all colimits. To define $F_!$, take any presheaf $P \in \mathbf{Sets}^{\mathbf{C}^{\mathrm{op}}}$ and write it as a canonical colimit of representables

$$\varinjlim_{j \in J} yC_j \cong P$$

with $J = \int_{\mathbf{C}} P$ the category of elements of P, as in proposition 8.10. Then, set

$$F_!(P) = \varinjlim_{j \in J} FC_j$$

with the colimit taken in $\mathcal{E}$, which is cocomplete. (We leave it to the reader to determine how to define $F_!$ on arrows.) Clearly, if $F_!$ is to preserve all colimits and satisfy (9.7), then up to isomorphism this must be its value for P. For F^*, take any $E \in \mathcal{E}$ and $C \in \mathbf{C}$ and observe that by (Yoneda and) the intended adjunction, for $F^*(E)(C)$, we must have

$$F^*(E)(C) \cong \mathrm{Hom}_{\hat{C}}(yC, F^*(E))$$
$$\cong \mathrm{Hom}_{\mathcal{E}}(F_!(yC), E)$$
$$\cong \mathrm{Hom}_{\mathcal{E}}(FC, E).$$

Thus, we simply set

$$F^*(E)(C) = \mathrm{Hom}_{\mathcal{E}}(FC, E)$$

which is plainly a presheaf on $\mathbf{C}$ (we use here that $\mathcal{E}$ is locally small). Now let us check that indeed $F_! \dashv F^*$. For any $E \in \mathcal{E}$ and $P \in \hat{\mathbf{C}}$, we have natural isomorphisms

$$\mathrm{Hom}_{\hat{\mathbf{C}}}(P, F^*(E)) \cong \mathrm{Hom}_{\hat{\mathbf{C}}}(\varinjlim_{j \in J} yC_j, F^*(E))$$

$$\cong \varprojlim_{j \in J} \mathrm{Hom}_{\hat{\mathbf{C}}}(yC_j, F^*(E))$$

$$\cong \varprojlim_{j \in J} F^*(E)(C_j)$$

$$\cong \varprojlim_{j \in J} \mathrm{Hom}_{\mathcal{E}}(FC_j, E)$$

$$\cong \mathrm{Hom}_{\mathcal{E}}(\varinjlim_{j \in J} FC_j, E)$$

$$\cong \mathrm{Hom}_{\mathcal{E}}(F_!(P), E).$$

Finally, for any object $C \in \mathbf{C}$,

$$F_!(yC) = \varinjlim_{j \in J} FC_j \cong FC$$

since the category of elements J of a representable yC has a terminal object, namely the element $1_C \in \mathrm{Hom}_{\mathbf{C}}(C, C)$. $\square$

Corollary 9.17. *Let* $f : \mathbf{C} \to \mathbf{D}$ *be a functor between small categories. The precomposition functor*

$$f^* : \mathbf{Sets}^{\mathbf{D}^{\mathrm{op}}} \to \mathbf{Sets}^{\mathbf{C}^{\mathrm{op}}}$$

given by

$$f^*(Q)(C) = Q(fC)$$

has both left and right adjoints

$$f_! \vdash f^* \vdash f_*$$

Moreover, there is a natural isomorphism

$$f_! \circ y_{\mathbf{C}} \cong y_{\mathbf{D}} \circ f$$

as indicated in the following diagram:

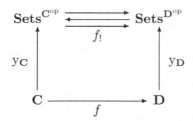

The induced functors $f_!$ and f_* are sometimes referred to in the literature as *(left and right) Kan extensions.*

Proof. First, define

$$F = y_{\mathbf{D}} \circ f : \mathbf{C} \to \mathbf{Sets}^{\mathbf{D}^{\mathrm{op}}}.$$

Then, by the foregoing proposition, we have adjoints $F_!$ and F^* as indicated in

$$
\begin{array}{ccc}
\mathbf{Sets}^{\mathbf{C}^{\mathrm{op}}} & \xleftarrow{\quad F^* \quad} & \mathbf{Sets}^{\mathbf{D}^{\mathrm{op}}} \\
& \xrightarrow{\quad F_! \quad} & \\
y_{\mathbf{C}} \uparrow & & \uparrow y_{\mathbf{D}} \\
\mathbf{C} & \xrightarrow{\quad f \quad} & \mathbf{D}
\end{array}
$$

and we know that $F_! \circ y_{\mathbf{C}} \cong y_D \circ f$. We claim that $F^* \cong f^*$. Indeed, by the definition of F^*, we have

$$F^*(Q)(C) = \mathrm{Hom}_{\hat{\mathbf{D}}}(FC, Q) \cong \mathrm{Hom}_{\hat{\mathbf{D}}}(y(fC), Q) \cong Q(fC) = f^*(Q)(C).$$

This, therefore, gives the functors $f_! \dashv f^*$. For f_*, apply the foregoing proposition to the composite

$$f^* \circ y_{\mathbf{D}} : \mathbf{D} \to \mathbf{Sets}^{\mathbf{D}^{\mathrm{op}}} \to \mathbf{Sets}^{\mathbf{C}^{\mathrm{op}}}.$$

This gives an adjunction

$$(f^* \circ y_{\mathbf{D}})_! \dashv (f^* \circ y_{\mathbf{D}})^*$$

so we just need to show that

$$(f^* \circ y_{\mathbf{D}})_! \cong f^*$$

in order to get the required right adjoint as $f_* = (f^* \circ y_{\mathbf{D}})^*$. By the universal property of $\mathbf{Sets}^{\mathbf{D}^{\mathrm{op}}}$, it suffices to show that f^* preserves colimits. But for any colimit $\varinjlim_j Q_j$ in $\mathbf{Sets}^{\mathbf{D}^{\mathrm{op}}}$

$$(f^*(\varinjlim_j Q_j))(C) \cong (\varinjlim_j Q_j)(fC)$$

$$\cong \varinjlim_j (Q_j(fC))$$

$$\cong \varinjlim_j ((f^* Q_j)(C))$$

$$\cong (\varinjlim_j (f^* Q_j))(C).$$

$\square$

This corollary says that, in a sense, *every functor has an adjoint!* For, given any $f : \mathbf{C} \to \mathbf{D}$, we indeed have the right adjoint

$$f^* \circ \mathbf{y_D} : \mathbf{D} \to \hat{\mathbf{C}}$$

except that its values are in the "ideal elements" of the cocompletion $\hat{\mathbf{C}} = \mathbf{Sets}^{\mathbf{C}^{\mathrm{op}}}$.

9.7 Locally cartesian closed categories

A special case of the situation described by corollary 9.17 is the *change of base* for indexed families of sets along a "reindexing" function $\alpha : J \to I$. An arbitrary such function between sets gives rise, by that corollary, to a triple of adjoint functors:

$$\mathbf{Sets}^J \begin{array}{c} \xrightarrow{\quad \alpha_* \quad} \\ \xleftarrow{\quad \alpha^* \quad} \\ \xrightarrow{\quad \alpha_! \quad} \end{array} \mathbf{Sets}^I$$

$$\alpha_! \dashv \alpha^* \dashv \alpha_*$$

Let us examine these functors more closely in this special case.

An object A of $\mathbf{Sets}^I$ is an I-indexed family of sets

$$(A_i)_{i \in I}.$$

Then, $\alpha^*(A) = A \circ \alpha$ is the reindexing of A along α to a J-indexed family of sets

$$\alpha^*(A) = (A_{\alpha(j)})_{j \in J}.$$

Given a J-indexed family B, let us calculate $\alpha_!(B)$ and $\alpha_*(B)$.

Consider first the case $I = 1$ and $\alpha =!_J : J \to 1$. Then, $(!_J)^* : \mathbf{Sets} \to \mathbf{Sets}^J$ is the "constant family" or diagonal functor $\Delta(A)(j) = A$, for which we know the adjoints:

$$\mathbf{Sets}^J \begin{array}{c} \xrightarrow{\quad \Pi \quad} \\ \xleftarrow{\quad \Delta \quad} \\ \xrightarrow{\quad \Sigma \quad} \end{array} \mathbf{Sets}$$

$$\Sigma \dashv \Delta \dashv \Pi$$

These are, namely, just the (disjoint) sum and cartesian product of the sets in the family

$$\sum_{j \in J} B_j, \qquad \prod_{j \in J} B_j.$$

Recall that we have the adjunctions:

$$\frac{\vartheta_j : B_j \to A}{(\vartheta_j) : \sum_j B_j \to A}, \qquad \frac{\vartheta_j : A \to B_j}{\langle \vartheta_j \rangle : A \to \prod_j B_j}$$

By uniqueness of adjoints, it therefore follows that $(!_J)_! \cong \Sigma$ and $(!_J)_* \cong \Pi$.

A general reindexing $\alpha : J \to I$ gives rise to generalized sum and product operations along α

$$\Sigma_\alpha \dashv \alpha^* \dashv \Pi_\alpha$$

defined on J-indexed families (B_j) by

$$(\Sigma_\alpha(B_j))_i = \sum_{\alpha(j)=i} B_j$$

$$(\Pi_\alpha(B_j))_i = \prod_{\alpha(j)=i} B_j.$$

These operations thus assign to an element $i \in I$ the sum, respectively the product, over all the sets indexed by the elements j in the preimage $\alpha^{-1}(i)$ of i under α.

Now let us recall from example 7.29 the equivalence between J-indexed families of sets and the slice category of "sets over J"

$$\mathbf{Sets}^J \simeq \mathbf{Sets}/J.$$

It takes a family $(A_j)_{j \in J}$ to the indexing projection $p : \sum_{j \in J} A_j \to J$ and a map $\pi : A \to J$ to the family $(\pi^{-1}(j))_{j \in J}$. We know, moreover, from an exercise in Chapter 7 that this equivalence respects reindexing, in the sense that for any $\alpha : J \to I$ the following square commutes up to natural isomorphism:

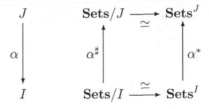

Here we write $\alpha^\sharp$ for the pullback functor along α. Since α^* has both right and left adjoints, we have the diagram of induced adjoints:

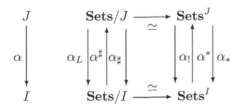

Proposition 9.18. *For any function $\alpha : J \to I$, the pullback functor $\alpha^\sharp :$* **Sets**$/I \to$ **Sets**$/J$ *has both left and right adjoints:*

$$\alpha_L \dashv \alpha^\sharp \dashv \alpha_\sharp$$

In particular, $\alpha^\sharp$ therefore preserves all limits and colimits.

Let us compute the functors explicitly. Given $\pi : A \to J$, let $A_j = \pi^{-1}(j)$ and recall that

$$\alpha_!(A)_i = \sum_{\alpha(j)=i} A_j.$$

But then, we have

$$
\begin{aligned}
\alpha_!(A)_i &= \sum_{\alpha(j)=i} A_j \\
&= \sum_{j \in \alpha^{-1}(i)} A_j \\
&= \sum_{j \in \alpha^{-1}(i)} \pi^{-1}(j) \\
&= \pi^{-1} \circ \alpha^{-1}(i) \\
&= (\alpha \circ \pi)^{-1}(i).
\end{aligned}
$$

It follows that $\alpha_L(\pi : A \to J)$ is simply the composite $\alpha \circ \pi : A \to J \to I$,

$$\alpha_L(\pi : A \to J) = (\alpha \circ \pi : A \to J \to I).$$

Indeed, the UMP of pullbacks essentially states that composition along any function α is left adjoint to pullback along α.

As for the right adjoint

$$\alpha_\sharp : \mathbf{Sets}/J \longrightarrow \mathbf{Sets}/I$$

given $\pi : A \to J$, the result $\alpha_\sharp(\pi) : \alpha_\sharp(A) \to I$ can be described fiberwise by

$$(\alpha_\sharp(A))_i = \{s : \alpha^{-1}(i) \to A \mid \text{``}s \text{ is a partial section of } \pi\text{''}\}$$

where the condition "s is a partial section of π" means that the following triangle commutes with the canonical inclusion $\alpha^{-1}(i) \subseteq J$ at the base.

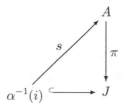

Henceforth, we also write these "change of base" adjoints along a map $\alpha :$ $J \to I$ in the form

$$
\begin{array}{ccc}
J & \mathbf{Sets}/J & \\
\alpha \downarrow & \Sigma_\alpha \downarrow \, \alpha^* \uparrow \, \Pi_\alpha \downarrow & \Sigma_\alpha \dashv \alpha^* \dashv \Pi_\alpha \\
I & \mathbf{Sets}/I &
\end{array}
$$

Finally, let us reconsider the case $I = 1$, where these adjoints take the form

$$
\begin{array}{ccc}
J & \mathbf{Sets}/J & \\
! \downarrow & \Sigma_J \downarrow \, J^* \uparrow \, \Pi_J \downarrow & \Sigma_J \dashv J^* \dashv \Pi_J \\
1 & \mathbf{Sets} &
\end{array}
$$

In this case, we have

$$
\Sigma_J(\pi : A \to J) \;=\; A
$$
$$
J^*(A) \;=\; (p_1 : J \times A \to J)
$$
$$
\Pi_J(\pi : A \to J) \;=\; \{ s : J \to A \mid \pi \circ s = 1 \}
$$

as the reader can easily verify. Moreover, one therefore has

$$
\Sigma_J J^*(A) \;=\; J \times A
$$
$$
\Pi_J J^*(A) \;=\; A^J.
$$

Thus, the product $\dashv$ exponential adjunction can be factored as a composite of adjunctions as follows:

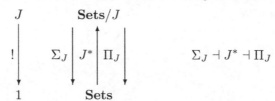

The following definition captures the notion of a category having this sort of adjoint structure. In such a category $\mathcal{E}$, the slice categories can be regarded as categories of abstract-indexed families of objects of $\mathcal{E}$, and the reindexing of such families can be carried out, with associated adjoint operations of sum and product.

Definition 9.19. A category $\mathcal{E}$ is called *locally cartesian closed* if $\mathcal{E}$ has a terminal object and for every arrow $f : A \to B$ in $\mathcal{E}$, the composition functor

$$\Sigma_f : \mathcal{E}/A \to \mathcal{E}/B$$

has a right adjoint f^* which, in turn, has a right adjoint Π_f:

$$\Sigma_f \dashv f^* \dashv \Pi_f$$

The choice of name for such categories is explained by the following important fact.

Proposition 9.20. *For any category $\mathcal{E}$ with a terminal object, the following are equivalent:*

  1. *$\mathcal{E}$ is locally cartesian closed.*

  2. *Every slice category $\mathcal{E}/A$ of $\mathcal{E}$ is cartesian closed.*

Proof. Let $\mathcal{E}$ be locally cartesian closed. Since $\mathcal{E}$ has a terminal object, products and exponentials in $\mathcal{E}$ can be built as

$$A \times B = \Sigma_B B^* A$$

$$B^A = \Pi_B B^* A.$$

Therefore, $\mathcal{E}$ is cartesian closed. But clearly every slice category $\mathcal{E}/X$ is also locally cartesian closed, since "a slice of a slice is a slice." Thus, every slice of $\mathcal{E}$ is cartesian closed.

Conversely, suppose every slice of $\mathcal{E}$ is cartesian closed. Then $\mathcal{E}$ has pullbacks, since these are just binary products in a slice. Thus, we just need to construct the "relative product" functor $\Pi_f : \mathcal{E}/A \to \mathcal{E}/B$ along a map $f : A \to B$. First, change notation:

$$\mathcal{F} = \mathcal{E}/B$$

$$F = f : A \to B$$

$$\mathcal{F}/F = \mathcal{E}/A$$

Thus, we want to construct $\Pi_F : \mathcal{F}/F \to \mathcal{F}$. Given an object $p : X \to F$ in $\mathcal{F}/F$, the object $\Pi_F(p)$ is constructed as the following pullback:

$$
\begin{array}{ccc}
\Pi_F(p) & \longrightarrow & X^F \\
\big\downarrow & & \big\downarrow {\scriptstyle p^F} \\
1 & \underset{\widetilde{1_F}}{\longrightarrow} & F^F
\end{array}
\qquad (9.8)
$$

where $\widetilde{1_F}$ is the exponential transpose of the composite arrow

$$1 \times F \cong F \xrightarrow{\ 1\ } F.$$

It is now easy to see from (9.8) that there is a natural bijection of the form

$$\frac{Y \to \Pi_F(p)}{F^*Y \to p}$$

$\square$

Remark 9.21. The reader should be aware that some authors do not require the existence of a terminal object in the definition of a locally cartesian closed category.

Example 9.22 (Presheaves). For any small category $\mathbf{C}$, the category $\mathbf{Sets}^{\mathbf{C}^{\mathrm{op}}}$ of presheaves on $\mathbf{C}$ is locally cartesian closed. This is a consequence of the following fact.

Lemma 9.23. *For any object* $P \in \mathbf{Sets}^{\mathbf{C}^{\mathrm{op}}}$, *there is a small category* $\mathbf{D}$ *and an equivalence of categories,*

$$\mathbf{Sets}^{\mathbf{C}^{\mathrm{op}}}/P \simeq \mathbf{Sets}^{\mathbf{D}^{\mathrm{op}}}.$$

Moreover, there is also a functor $p : \mathbf{D} \to \mathbf{C}$ *such that the following diagram commutes (up to natural isomorphism):*

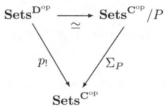

Proof. One can take

$$\mathbf{D} = \int_{\mathbf{C}} P$$

$$p = \pi : \int_{\mathbf{C}} P \to \mathbf{C}$$

Indeed, recall that by the Yoneda lemma, the category $\int_{\mathbf{C}} P$ of elements of P can be described equivalently (isomorphically, in fact) as the category that we write suggestively as y/P, described as follows:

Objects: pairs (C, x) where $C \in \mathbf{C}$ and $x : \mathrm{y}C \to P$ in $\mathbf{Sets}^{\mathbf{C}^{\mathrm{op}}}$

Arrows: all arrows between such objects in the slice category over P

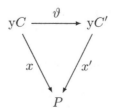

Note that by Yoneda, each such arrow is of the form $\vartheta = \mathrm{y}h$ for a unique $h : C \to D$ in $\mathbf{C}$, which, moreover, is such that $P(h)(x') = x$.

Now let $I : \mathrm{y}/P \to \mathbf{Sets}^{\mathbf{C}^{\mathrm{op}}}/P$ be the evident (full and faithful) inclusion functor, and define a functor

$$\Phi : \mathbf{Sets}^{\mathbf{C}^{\mathrm{op}}}/P \to \mathbf{Sets}^{(\mathrm{y}/P)^{\mathrm{op}}}$$

by setting, for any $q : Q \to P$ and $(C, x) \in \mathrm{y}/P$

$$\Phi(q)(C, x) = \mathrm{Hom}_{\hat{\mathbf{C}}/P}(x, q),$$

the elements of which look like

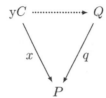

In other words, $\Phi(q) = I^*(\mathrm{y}q)$, which is plainly functorial. We leave it to the reader as an exercise to show that this functor establishes an equivalence of categories. $\qquad\Box$

Combining the foregoing with the fact (theorem 8.14) that categories of presheaves are always cartesian closed now yields the promised:

Corollary 9.24. *For any small category* $\mathbf{C}$*, the category* $\mathbf{Sets}^{\mathbf{C}^{\mathrm{op}}}$ *of presheaves on* $\mathbf{C}$ *is locally cartesian closed.*

Remark 9.25. Part of the interest in locally cartesian closed categories derives from their use in the semantics of dependent type theory, which has type-indexed families of types

$$x : A \vdash B(x)$$

and type constructors of dependent sum and product

$$\sum_{x:A} B(x) \qquad \prod_{x:A} B(x).$$

Indeed, just as cartesian closed categories provide a categorical interpretation of the simply typed λ-calculus, so locally cartesian closed categories interpret the dependently typed λ-calculus. And since the Yoneda embedding preserves CCC structure, the completeness theorem for λ-calculus with respect to arbitrary CCCs (theorem 6.17) implies completeness with respect to just categories of presheaves $\mathbf{Sets}^{\mathbf{C}^{\mathrm{op}}}$, as was shown in exercise 10 of Chapter 8. Now, just the same sort of completeness theorem holds for dependent type theory as well, by an elementary argument involving the foregoing lemma. More difficult to prove is the fact that one can do even better, retaining completeness while restricting the interpretations to just the "categories of diagrams" on *posets*, $\mathbf{Sets}^{\mathrm{P}}$, which can be regarded as Kripke models (and this of course then also holds for the simply typed λ-calculus as well). In this connection, the following alternate description of such categories is then of particular interest.

Example 9.26 Fibrations of posets. A monotone map of posets $f : X \to P$ is a (discrete) *fibration* if it has the following *lifting property*:

For every $x \in X$ and $p' \leq fx$, there is a unique $x' \leq x$ such that $f(x') = p'$.

One says that x "lies over" $p = f(x)$ and that any $p' \leq p$ "lifts" to a unique $x' \leq x$ lying over it, as indicated in the following diagram:

$$
\begin{array}{ccc}
X & & x' \cdots\cdots\cdots\cdots\cdots\to x \\
\Big\downarrow f & & \qquad\leq \\
P & & p' \xrightarrow{\quad\leq\quad} p
\end{array}
$$

The identity morphism of a given poset P is clearly a fibration, and the composite of two fibrations is easily seen to be a fibration. Let $\mathbf{Fib}$ denote the (non-full) subcategory of posets and fibrations between them as arrows.

Lemma 9.27. *For any poset P, the slice category $\mathbf{Fib}/P$ is cartesian closed.*

Proof. The category $\mathbf{Fib}/P$ is equivalent to the category of presheaves on P,

$$
\mathbf{Fib}/P \simeq \mathbf{Sets}^{P^{\mathrm{op}}}.
$$

To get a functor, $\Phi : \mathbf{Fib}/P \to \mathbf{Sets}^{P^{\mathrm{op}}}$, takes a fibration $q : Q \to P$ to the presheaf defined on objects by

$$
\Phi(q)(p) = q^{-1}(p) \qquad \text{for } p \in P.
$$

The lifting property then determines the action on arrows $p' \leq p$. For the other direction, $\Psi : \mathbf{Sets}^{P^{\mathrm{op}}} \to \mathbf{Fib}/P$ takes a presheaf $Q : P^{\mathrm{op}} \to \mathbf{Sets}$ to (the indexing projection of) its category of elements,

$$
\Psi(Q) = \int_P Q \xrightarrow{\ \pi\ } P.
$$

These are easily seen to be quasi-inverses. $\qquad\qquad\qquad\qquad\qquad\square$

The category $\mathbf{Fib}$ itself is *almost* locally cartesian closed; it only lacks a terminal object (why?). We can "fix" this simply by slicing it.

Corollary 9.28. *For any poset P, the slice category $\mathbf{Fib}/P$ is locally cartesian closed.*

This sort of case is not uncommon, which is why the notion "locally cartesian closed" is sometimes formulated without requiring a terminal object.

9.8 Adjoint functor theorem

The question we now want to consider systematically is, when does a functor have an adjoint? Consider first the question, when does a functor of the form $\mathbf{C} \to \mathbf{Sets}$ have a left adjoint? If $U : \mathbf{C} \to \mathbf{Sets}$ has $F \dashv U$, then U is representable $U \cong \mathrm{Hom}(F1, -)$, since $U(C) \cong \mathrm{Hom}(1, UC) \cong \mathrm{Hom}(F1, C)$.

A related condition that makes sense for categories other than $\mathbf{Sets}$ is preservation of limits. Suppose that $\mathbf{C}$ is complete and $U : \mathbf{C} \to \mathbf{X}$ preserves limits; then we can ask whether U has a left adjoint. The *adjoint functor theorem* (AFT) gives a necessary and sufficient condition for this case.

Theorem 9.29 (Freyd). *Let $\mathbf{C}$ be locally small and complete. Given any category $\mathbf{X}$ and a limit-preserving functor*

$$U : \mathbf{C} \to \mathbf{X}$$

the following are equivalent:

1. *U has a left adjoint.*
2. *For each object $X \in \mathbf{X}$, the functor U satisfies the following:*
   Solution set condition: There exists a set of objects $(S_i)_{i \in I}$ in $\mathbf{C}$ such that for any object $C \in \mathbf{C}$ and arrow $f : X \to UC$, there exists an $i \in I$ and arrows $\varphi : X \to US_i$ and $\bar{f} : S_i \to C$ such that

$$f = U(\bar{f}) \circ \varphi$$

Briefly: "every arrow $X \to UC$ factors through some object S_i in the solution set."

For the proof, we require the following.

Lemma 9.30. *Let $\mathbf{D}$ be locally small and complete. Then the following are equivalent:*

1. *$\mathbf{D}$ has an initial object.*
2. *$\mathbf{D}$ satisfies the following:*

*Solution set condition: There is a set of objects $(D_i)_{i \in I}$ in **D** such that for any object $D \in$ **C**, there is an arrow $D_i \to D$ for some $i \in I$.*

Proof. If **D** has an initial object 0, then $\{0\}$ is obviously a solution set.

Conversely, suppose we have a solution set $(D_i)_{i \in I}$ and consider the object

$$W = \prod_{i \in I} D_i,$$

which exists since I is small and **D** is complete. Now W is "weakly initial" in the sense that for any object D there is a (not necessarily unique) arrow $W \to D$, namely the composite

$$\prod_{i \in I} D_i \to D_i \to D$$

for a suitable product projection $\prod_{i \in I} D_i \to D_i$. Next, take the joint equalizer of all endomorphisms $d : W \to W$ (which is a set, since **D** is locally small), as indicated in the diagram:

$$V \xrightarrow{\ h\ } W \underset{\langle d \rangle}{\overset{\Delta}{\rightrightarrows}} \prod_{d:W \to W} W$$

Here, the arrows Δ and $\langle d \rangle$ have the d-projections $1_W : W \to W$ and $d : W \to W$, respectively. This equalizer then has the property that for any endomorphism $d : W \to W$,

$$d \circ h = h. \tag{9.9}$$

Note, moreover, that V is still weakly initial, since for any D there is an arrow $V \rightarrowtail W \to D$. Suppose that for some D there are two arrows $f, g : V \to D$. Take their equalizer $e : U \to V$, and consider the following diagram:

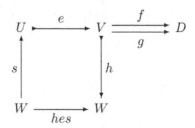

in which the arrow s comes from W being weakly initial. So for the endomorphism hes by (9.9), we have

$$hesh = h.$$

Since h is monic, $esh = 1_V$. But then $eshe = e$, and so also $she = 1_U$ since e is monic. Therefore $U \cong V$, and so $f = g$. Thus, V is an initial object. $\qquad \square$

Now we can prove the theorem.

Proof. (Theorem) If U has a left adjoint $F \dashv U$, then $\{FX\}$ is itself a solution set for X, since we always have a factorization,

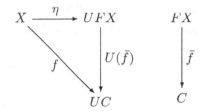

where $\bar{f} : FX \to C$ is the adjoint transpose of f and $\eta : X \to UFX$ the unit of the adjunction.

Conversely, consider the following so-called *comma-category* $(X|U)$, with

Objects: are pairs (C, f) with $f : X \to UC$

Arrows: $g : (C, f) \to (C', f')$ are arrows $g : C \to C'$ with $f' = U(g)f$.

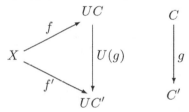

Clearly, U has a left adjoint F iff for each object X this category $(X|U)$ has an initial object, $(FX, \eta : X \to UFX)$, which then has the UMP of the unit. Thus, to use the foregoing initial object lemma, we must check

1. $(X|U)$ is locally small.
2. $(X|U)$ satisfies the solution set condition in the lemma.
3. $(X|U)$ is complete.

For (1), we just observe that **C** is locally small. For (2), the solution set condition of the theorem implies that there is a set of objects,

$$\{(S_i, \varphi : X \to US_i) \mid i \in I\}$$

such that every object $(C, f : X \to UC)$ has an arrow $\bar{f} : (S_i, \varphi) \to (C, f)$.

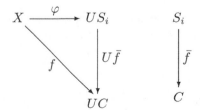

Finally, to see that $(X|U)$ is complete, one can easily check directly that it has products and equalizers, using the fact that $\mathbf{C}$ has and U preserves these. We leave this as an easy exercise for the reader. $\square$

Remark 9.31. 1. The theorem simply does not apply if $\mathbf{C}$ is not complete. In that case, a given functor may have an adjoint, but the AFT will not tell us that.

   2. It is essential that the solution set in the theorem be a *set* (and that $\mathbf{C}$ have all set-sized limits).

   3. On the other hand, if $\mathbf{C}$ is itself *small* and complete, then we can plainly drop the solution set condition entirely. In that case, we have the following.

Corollary 9.32. *If* $\mathbf{C}$ *is a small and complete category and* $U : \mathbf{C} \to \mathbf{X}$ *is a functor that preserves all limits, then* U *has a left adjoint.*

Example 9.33. For complete posets P, Q, a monotone function $f : P \to Q$ has a right adjoint $g : Q \to P$ iff f is *cocontinuous*, in the sense that $f(\bigvee_i p_i) = \bigvee_i f(p_i)$ for any set-indexed family of elements $(p_i)_{i \in I}$. (Of course, here we are using the dual formulation of the AFT.)

Indeed, we can let

$$g(q) = \bigvee_{f(x) \leq q} x.$$

Then for any $p \in P$ and $q \in Q$, if

$$p \leq g(q)$$

then

$$f(p) \leq fg(q) = f(\bigvee_{f(x) \leq q} x) = \bigvee_{f(x) \leq q} f(x) \leq q.$$

While, conversely, if

$$f(p) \leq q$$

then clearly

$$p \leq \bigvee_{f(x) \leq q} x = g(q).$$

As a further consequence of the AFT, we have the following characterization of representable functors on small complete categories.

Corollary 9.34. *If* $\mathbf{C}$ *is a small and complete category, then for any functor* $U : \mathbf{C} \to \mathbf{Sets}$ *the following are equivalent:*

   1. U preserves all limits.

   2. U has a left adjoint.

   3. U is representable.

Proof. Immediate. □

These corollaries are, however, somewhat weaker than it may at first appear, in light of the following fact.

Proposition 9.35. *If* **C** *is small and complete, then* **C** *is a preorder.*

Proof. Suppose not, and take $C, D \in \mathbf{C}$ with $\mathrm{Hom}(C, D) \geq 2$. Let J be any set, and take the product

$$\prod_J D.$$

There are isomorphisms:

$$\mathrm{Hom}(C, \prod_J D) \cong \prod_J \mathrm{Hom}(C, D) \cong \mathrm{Hom}(C, D)^J$$

So, for the cardinalities of these sets, we have

$$|\mathrm{Hom}(C, \prod_J D)| = |\mathrm{Hom}(C, D)|^{|J|} \geq 2^{|J|} = |P(J)|.$$

And that is for *any* set J. On the other hand, clearly $|\mathbf{C}_1| \geq |\mathrm{Hom}(C, \prod_J D)|$. So taking $J = \mathbf{C}_1$ in the above calculation gives a contradiction. □

Remark 9.36. An important special case of the AFT that often occurs "in nature" is that in which the domain category satisfies certain conditions that eliminate the need for the (rather unpleasant!) solution set condition entirely. Specifically, let **A** be a locally small, complete category satisfying the following conditions:

1. **A** is *well powered*: each object A has at most a *set* of subobjects $S \rightarrowtail A$.

2. **A** has a *cogenerating set*: there is a set of objects $\{A_i \mid i \in I\}$ (*I* some index *set*), such that for any A, X and $x \neq y : X \rightrightarrows A$ in **A**, there is some $s : A \to A_i$ (for some i) that "separates" x and y, in the sense that $sx \neq sy$.

Then any functor $U : \mathbf{A} \to \mathbf{X}$ that preserves limits necessarily has a left adjoint. In this form (also originally proved by Freyd), the theorem is usually known as the special adjoint functor theorem ("SAFT"). We refer to Mac Lane, V.8 for the proof, and some sample applications.

Example 9.37. An important application of the AFT is that any *equational theory* T gives rise to a free ⊣ forgetful adjunction between **Sets** and the category of models of the theory, or "T-algebras." In somewhat more detail, let T be a (finitary) equational theory, consisting of finitely many operation symbols, each of some finite arity (including nullary operations, i.e., constant symbols), and a set of equations between terms built from these operations and variables. For instance, the theory of groups has a constant u (the group unit), a unary operation g^{-1} (the inverse), and a binary operation $g \cdot h$ (the group product), and a handful of equations such as $g \cdot u = g$. The theory of rings has a further binary operation and some more equations. The theory of fields is not equational, however, because the condition $x \neq 0$ is required for an element x to have a multiplicative inverse. A T-algebra is a set equipped with operations (of the proper arities) corresponding to the operation symbols in T, and satisfying the equations of T. A homomorphism of T-algebras $h : A \to B$ is a function on the underlying sets that preserves all the operations, in the usual sense. Let T-**Alg** be the category of all such algebras and their homomorphisms. There is an evident forgetful functor

$$U : T\text{-}\mathbf{Alg} \to \mathbf{Sets}.$$

The AFT implies that this functor always has a left adjoint F, the "free algebra" functor.

Proposition 9.38. *For any equational theory T, the forgetful functor from T-algebras to* **Sets** *has a left adjoint.*

Rather than proving this general proposition (for which see Mac Lane, chapter V), it is more illuminating to do a simple example.

Example 9.39. Let T be the theory with one constant and one unary operation (no axioms). A T-algebra is a set M with the structure

$$1 \xrightarrow{a} M \xrightarrow{f} M$$

If $1 \xrightarrow{b} N \xrightarrow{g} N$ is another such algebra, a homomorphism of T-algebras $\phi : (M, a, f) \to (N, b, g)$ is a function $\phi : M \to N$ that preserves the element and the operation, in the expected sense that

$$\phi a = b$$
$$\phi f = g\phi.$$

as indicated in the commutative diagram:

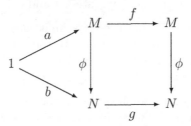

There is an evident forgetful functor (forget the T-algebra structure):

$$U : T\text{-}\mathbf{Alg} \to \mathbf{Sets}.$$

This functor is easily seen to create all limits, as is the case for algebras for any theory T. So in particular, T-**Alg** is complete and U preserves limits. Thus in order to apply the AFT, we just need to check the solution set condition.

To that end, let X be any set and take any function

$$h : X \to M.$$

The image $h(X) \subseteq M$ generates a sub-T-model of (M, a, f) as follows. Define the set "generated by $h(X)$" to be

$$H = \langle h(X) \rangle = \{ f^n(z) \mid n \in \mathbb{N}, z = a \text{ or } z = h(x) \text{ for some } x \in X \}. \quad (9.10)$$

Then $a \in H$, and f restricts to H to give a function $f' : H \to H$. Moreover, the inclusion $i : H \hookrightarrow M$ is clearly a T-algebra homomorphism

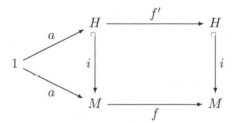

Furthermore, since $h(X) \subseteq H$ there is a factorization h' of h, as indicated in the following diagram:

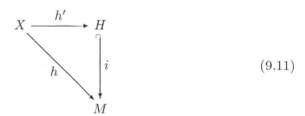

$$(9.11)$$

Now observe that, given X, the cardinality $|H|$ is *bounded*, that is, for a sufficiently large κ independent of h and M, we have

$$|H| \leq \kappa.$$

Indeed, inspecting (9.10), we can take $\kappa = |\mathbb{N}| \times (1 + |X|)$.

To find a solution set for X, let us now take one representative N of each isomorphism class of T-algebras with cardinality at most κ. The set of all such algebras N is then a solution set for X and U. Indeed, as we just showed, any function $h : X \to M$ factors as in (9.11) through an element of this set (namely an isomorphic copy N of H). By the AFT, there thus exists a free functor,

$$F : \mathbf{Sets} \to T\text{-}\mathbf{Alg}.$$

A precisely analogous argument works for any equational theory T.

Finally, let us consider the particular free model $F(\emptyset)$ in T-**Alg**. Since left adjoints preserve colimits, this is an initial object. It follows that $F(\emptyset)$ is a *natural numbers object*, in the following sense.

Definition 9.40. Let **C** be a category with a terminal object 1. A natural numbers object (NNO) in **C** is a structure of the form

$$1 \xrightarrow{0} N \xrightarrow{s} N$$

which is initial among all such structures. Precisely, given any $1 \xrightarrow{a} X \xrightarrow{f} X$ in **C**, there is a unique arrow $\phi : N \to X$ such that the following commutes:

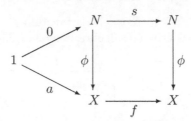

In other words, given any object X, a "starting point" $a \in X$ and an operation $x \mapsto f(x)$ on X, we can build up a unique $\phi : N \to X$ *recursively* by the equations:

$$\phi(0) = a$$

$$\phi(s(n)) = f(\phi(n)) \quad \text{for all } n \in N$$

Thus, the UMP of an NNO says precisely that such an object supports recursive definitions. It is easy to show that the set $\mathbb{N}$ of natural numbers with the canonical structure of 0 and the "successor function" $s(n) = n + 1$ is an NNO, and thus, by the UMP any NNO in **Sets** is isomorphic to it. The characterization of $\mathbb{N}$ in terms of the UMP of recursive definitions is therefore equivalent to the usual logical definition using the Peano axioms in **Sets**. But note that the notion of an NNO (which is due to F.W. Lawvere) also makes sense in many categories where the Peano axioms do not make any sense, since the latter involve logical operations like quantifiers.

Let us consider some simple examples of recursively defined functions using this UMP.

Example 9.41.   1. Let $(N, 0, s)$ be an NNO in any category **C**. Take any point $a : 1 \to N$, and consider the new structure:

$$1 \xrightarrow{a} N \xrightarrow{s} N$$

Then by the universal property of the NNO, there is a unique morphism $f_a : N \to N$ such that the following commutes:

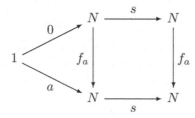

ADJOINTS   247

Thus we have the following "recursion equations":

$$f_a(0) = a$$
$$f_a(s(n)) = s(f_a(n))$$

If we write $f_a(n) = a + n$, then the above equations become the familiar recursive definition of addition:

$$a + 0 = a$$
$$a + (sn) = s(a + n)$$

2. Now take this arrow $a + (-) : N \to N$ together with $0 : 1 \to N$ to get another arrow $g_a : N \to N$, which is the unique one making the following commute:

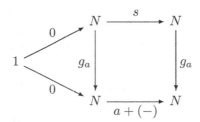

We then have the recursion equations:

$$g_a(0) = 0$$
$$g_a(sn) = a + g_a(n)$$

So, writing $g_a(n) = a \cdot n$, the above equations become the familiar recursive definition of multiplication:

$$a \cdot 0 = 0$$
$$a \cdot (sn) = a + a \cdot n$$

3. For an example of a different sort, suppose we have a (small) category $\mathbf{C}$ and an endofunctor $F : \mathbf{C} \to \mathbf{C}$. Then there is a structure

$$1 \xrightarrow{\text{id}} \mathbf{C}^{\mathbf{C}} \xrightarrow{F^{\mathbf{C}}} \mathbf{C}^{\mathbf{C}}$$

where $\text{id} : 1 \to \mathbf{C}^{\mathbf{C}}$ is the transpose of the identity $1_{\mathbf{C}} : \mathbf{C} \to \mathbf{C}$ (composed with the iso projection $1 \times \mathbf{C} \cong \mathbf{C}$). We therefore have a unique functor $f : \mathbb{N} \to \mathbf{C}^{\mathbf{C}}$ making the following diagram commute (we use the easy fact, which the reader should check, that the discrete category $\mathbb{N}$ is an NNO in **Cat**):

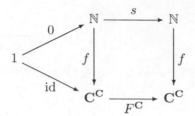

Transposing gives the commutative diagram

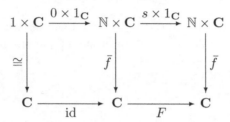

from which we can read off the recursion equations:

$$\bar{f}(0, C) = C$$
$$\bar{f}(sn, C) = F(\bar{f}(n, C))$$

It follows that $\bar{f}(n, C) = F^{(n)}(C)$, that is, $f(n)$ is the nth iterate of the functor $F : \mathbf{C} \to \mathbf{C}$.

9.9   Exercises

1. Complete the proof that the "Hom-set" definition of adjunction is equivalent to the preliminary one by showing that the specification of the unit $\eta_C : C \to UFC$ as $\eta_C = \phi(1_{FC})$ really is a natural transformation.

2. Show that every monoid M admits a surjection from a free monoid $F(X) \to M$, by considering the counit of the *free* ⊣ *forgetful* adjunction.

3. What is the unit of the product ⊣ exponential adjunction (say, in **Sets**)?

4. Let 2 be any two-element set and consider the "diagonal functor"

$$\Delta : \mathbf{C} \to \mathbf{C}^2$$

for any category **C**, that is, the exponential transpose of the first product projection

$$\mathbf{C} \times 2 \to \mathbf{C}.$$

Show that Δ has a right (resp. left) adjoint if and only if **C** has binary products (resp. coproducts).

Now let $\mathbf{C} = \mathbf{Sets}$ and replace 2 with an arbitrary small category $\mathbf{J}$. Determine both left and right adjoints for $\Delta : \mathbf{Sets} \to \mathbf{Sets^J}$. (Hint: $\mathbf{Sets}$ is complete and cocomplete.)

5. Let $\mathbf{C}$ be cartesian closed and suppose moreover that $\mathbf{C}$ has all finite colimits. Show that $\mathbf{C}$ is not only distributive,

$$(A + B) \times C \cong (A \times C) + (B \times C)$$

but that also $(-) \times C$ preserves coequalizers. Dually, show that $(-)^C$ preserves products and equalizers.

6. Any category $\mathbf{C}$ determines a preorder $P(\mathbf{C})$ by setting: $A \leq B$ if and only if there is an arrow $A \to B$. Show that the functor P is (left? right?) adjoint to the evident inclusion functor of preorders into categories. Does the inclusion also have an adjoint on the other side?

7. Show that there is a string of four adjoints between $\mathbf{Cat}$ and $\mathbf{Sets}$,

$$V \dashv F \dashv U \dashv R$$

where $U : \mathbf{Cat} \to \mathbf{Sets}$ is the forgetful functor to the set of objects $U(\mathbf{C}) = \mathbf{C}_0$. (Hint: for V, consider the "connected components" of a category.)

8. Given a function $f : A \to B$ between sets, verify that the direct image operation $\mathrm{im}(f) : P(A) \to P(B)$ is left adjoint to the inverse image $f^{-1} : P(B) \to P(A)$. Determine the dual image $f_* : P(A) \to P(B)$ and show that it is right adjoint to f^{-1}.

9. Show that the contravariant powerset functor $\mathcal{P} : \mathbf{Sets^{op}} \to \mathbf{Sets}$ is self-adjoint.

10. Given an object C in a category $\mathbf{C}$ under what conditions does the evident forgetful functor from the slice category $\mathbf{C}/C$

$$U : \mathbf{C}/C \to \mathbf{C}$$

have a right adjoint? What about a left adjoint?

11. (a) A *coHeyting algebra* is a poset P such that P^{op} is a Heyting algebra. Determine the coHeyting implication operation a/b in a lattice L by adjointness (with respect to joins), and show that any Boolean algebra is a coHeyting algebra by explicitly defining this operation a/b in terms of the usual Boolean ones.

(b) In a coHeyting algebra, there are operations of *coHeyting negation* $\sim p = 1/p$ and *coHeyting boundary* $\partial p = p \wedge \sim p$. State the logical rules of inference for these operations.

(c) A *biHeyting algebra* is a lattice that is both Heyting and coHeyting. Give an example of a biHeyting algebra that is not Boolean. (Hint: consider the lower sets in a poset.)

12. Let $\mathbf{P}$ be the category of propositions (i.e., the preorder category associated to the propositional calculus, say with countably many propositional

variables $p, q, r, \ldots$, and a unique arrow $p \to q$ if and only if $p \vdash q$). Show that for any fixed object p, there is a functor

$$- \wedge p : \mathbf{P} \to \mathbf{P}$$

and that this functor has a right adjoint. What is the counit of the adjunction? (When) does $- \wedge p$ have a left adjoint?

13. (a) Given any set I, explicitly describe the Yoneda embedding $y : I \to \mathbf{Sets}^I$ of I into the category $\mathbf{Sets}^I$ of I-indexed sets.

  (b) Given any function $f : J \to I$ from another set J, prove directly that the following diagram commutes up to natural isomorphism.

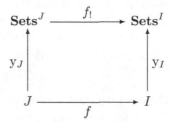

  (c) Describe the result of composing the Yoneda embedding with the equivalence,

$$\mathbf{Sets}^I \simeq \mathbf{Sets}/I.$$

  (d) What does the commutativity of the above "change of base" square mean in terms of the categories $\mathbf{Sets}/I$ and $\mathbf{Sets}/J$?

  (e) Consider the inclusion functor $i : \mathcal{P}(I) \to \mathbf{Sets}/I$ that takes a subset $U \subseteq I$ to its inclusion function $i(U) : U \to I$. Show that this is a functor and that it has a left adjoint

$$\sigma : \mathbf{Sets}/I \longrightarrow \mathcal{P}(I).$$

  (f) (Lawvere's Hyperdoctrine Diagram) In $\mathbf{Sets}$, given any function $f : I \to J$, consider the following diagram of functors:

$$
\begin{array}{ccc}
 & \overset{\Pi_f}{\longrightarrow} & \\
\mathbf{Sets}/I & \overset{f^*}{\longleftarrow} & \mathbf{Sets}/J \\
 & \underset{\Sigma_f}{\longrightarrow} & \\
\sigma_I \downarrow \uparrow i_I & & \sigma_J \downarrow \uparrow i_J \\
 & \overset{\forall_f}{\longrightarrow} & \\
\mathcal{P}(I) & \overset{f^{-1}}{\longleftarrow} & \mathcal{P}(J) \\
 & \underset{\exists_f}{\longrightarrow} &
\end{array}
$$

There are adjunctions $\sigma \dashv i$ (for both I and J), as well as $\Sigma_f \dashv f^* \dashv \Pi_f$ and $\exists_f \dashv f^{-1} \dashv \forall_f$, where $f^* : \mathbf{Sets}/J \to \mathbf{Sets}/I$ is pullback and $f^{-1} : \mathcal{P}(J) \to \mathcal{P}(I)$ is inverse image.

Consider which of the many possible squares commute. (Hint: first prove that a diagram of left adjoints commutes up to isomorphism if and only if the corresponding diagram consisting of their right adjoints does so.)

14. Complete the proof in the text that every slice of a category of presheaves is again a category of presheaves: for any small category $\mathbf{C}$ and presheaf $P : \mathbf{C}^{\mathrm{op}} \to \mathbf{Sets}$,
$$\mathbf{Sets}^{\mathbf{C}^{\mathrm{op}}}/P \simeq \mathbf{Sets}^{(\int_{\mathbf{C}} P)^{\mathrm{op}}}.$$

15. Let $\mathbf{C}$ be a complete category and $U : \mathbf{C} \to \mathbf{X}$ a continuous functor. Show that for any object $X \in \mathbf{X}$, the comma category $(X|U)$ is also complete.

16. Use the adjoint functor theorem to prove the following facts, which were shown by explicit constructions in Chapter 1:

    (a) Free monoids on sets exist.

    (b) Free categories on graphs exist.

17. Let $1 \xrightarrow{0} N \xrightarrow{s} N$ be an NNO in a cartesian closed category.

    (a) Show how to define the exponentiation operation m^n as an arrow $N \times N \to N$.

    (b) Do the same for the factorial function $n!$.

18. (Freyd's characterization of NNOs) Let $1 \xrightarrow{0} N \xrightarrow{s} N$ be an NNO in $\mathbf{Sets}$ (for your information, however, the following holds in any topos).

    (a) Prove that the following is a coproduct diagram:
    $$1 \xrightarrow{\quad 0 \quad} N \xleftarrow{\quad s \quad} N$$
    So $N \cong 1 + N$.

    (b) Prove that the following is a coequalizer:
    $$N \underset{1_N}{\overset{s}{\rightrightarrows}} N \longrightarrow 1$$

    (a) Show that any structure $1 \xrightarrow{0} N \xrightarrow{s} N$ satisfying the foregoing two conditions is an NNO.

19. Recall (from Chapter 1) the category $\mathbf{Rel}$ of relations (between sets), with arrows $R : A \to B$ being the relations $R \subseteq A \times B$ in $\mathbf{Sets}$. Taking the graph of a function $f : A \to B$ gives a relation $\Gamma(f) = \{(a, f(a)) \mid a \in A\} \subseteq A \times B$, and this assignment determines a functor $\Gamma : \mathbf{Sets} \to \mathbf{Rel}$. Show that Γ has a right adjoint. Compute the unit and counit of the adjunction.

MONADS AND ALGEBRAS

In Chapter 9, the adjoint functor theorem was seen to imply that the category of algebras for an equational theory T always has a "free T-algebra" functor, left adjoint to the forgetful functor into **Sets**. This adjunction describes the notion of a T-algebra in a way that is independent of the specific syntactic description given by the theory T, the operations and equations of which are rather like a particular *presentation* of that notion. In a certain sense that we are about to make precise, it turns out that *every* adjunction describes, in a "syntax invariant" way, a notion of an "algebra" for an abstract "equational theory."

Toward this end, we begin with yet a third characterization of adjunctions. This one has the virtue of being entirely equational.

10.1 The triangle identities

Suppose we are given an adjunction,

$$F : \mathbf{C} \rightleftarrows \mathbf{D} : U.$$

with unit and counit,

$$\eta : 1_{\mathbf{C}} \to UF$$

$$\epsilon : FU \to 1_{\mathbf{D}}.$$

We can take any $f : FC \to D$ to

$$\phi(f) = U(f) \circ \eta_C : C \to UD,$$

and for any $g : C \to UD$, we have

$$\phi^{-1}(g) = \epsilon_D \circ F(g) : FC \to D.$$

This we know gives the isomorphism

$$\mathrm{Hom}_{\mathbf{D}}(FC, D) \cong_\phi \mathrm{Hom}_{\mathbf{C}}(C, UD).$$

Now put $1_{UD} : UD \to UD$ in place of $g : C \to UD$ in the foregoing consideration. We know that $\phi^{-1}(1_{UD}) = \epsilon_D$, and so

$$1_{UD} = \phi(\epsilon_D)$$

$$= U(\epsilon_D) \circ \eta_{UD}.$$

And similarly, $\phi(1_{FC}) = \eta_C$, so

$$1_{FC} = \phi^{-1}(\eta_C)$$
$$= \epsilon_{FC} \circ F(\eta_C).$$

Thus, we have shown that the following two diagrams commute:

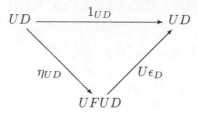

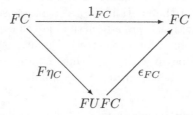

Indeed, one has the following equations of natural transformations:

$$U\epsilon \circ \eta_U = 1_U \qquad (10.1)$$
$$\epsilon_F \circ F\eta = 1_F \qquad (10.2)$$

These are called the "triangle identities."

Proposition 10.1. *Given categories, functors, and natural transformations*

$$F : \mathbf{C} \rightleftarrows \mathbf{D} : U$$

$$\eta : 1_{\mathbf{C}} \to U \circ F$$
$$\epsilon : F \circ U \to 1_{\mathbf{D}}$$

one has $F \dashv U$ with unit η and counit ϵ iff the triangle identities (10.1) *and* (10.2) *hold.*

Proof. We have already shown one direction. For the other, we just need a natural isomorphism,

$$\phi : \mathrm{Hom}_{\mathbf{D}}(FC, D) \cong \mathrm{Hom}_{\mathbf{C}}(C, UD).$$

As earlier, we put

$$\phi(f : FC \to D) = U(f) \circ \eta_C$$
$$\vartheta(g : C \to UD) = \epsilon_D \circ F(g).$$

Then we check that these are mutually inverse:

$$\phi(\vartheta(g)) = \phi(\epsilon_D \circ F(g))$$
$$= U(\epsilon_D) \circ UF(g) \circ \eta_C$$
$$= U(\epsilon_D) \circ \eta_{UD} \circ g \qquad\qquad \eta \text{ natural}$$
$$= g \qquad\qquad (10.1)$$

Similarly,

$$\vartheta(\phi(f)) = \vartheta(U(f) \circ \eta_C)$$
$$= \epsilon_D \circ FU(f) \circ F\eta_C$$
$$= f \circ \epsilon_{FC} \circ F\eta_C \qquad\qquad \epsilon \text{ natural}$$
$$= f \qquad\qquad (10.2)$$

Moreover, this isomorphism is easily seen to be natural. $\qquad\qquad\square$

The triangle identities have the virtue of being entirely "algebraic"—no quantifiers, limits, Hom-sets, infinite conditions, etc. Thus, anything defined by adjoints such as free groups, product spaces, quantifiers, ... can be defined *equationally*. This is not only a matter of conceptual simplification; it also has important consequences for the existence and properties of the structures that are so determined.

10.2 Monads and adjoints

Next consider an adjunction $F \dashv U$ and the composite functor
$$U \circ F : \mathbf{C} \to \mathbf{D} \to \mathbf{C}.$$
Given *any* category $\mathbf{C}$ and endofunctor
$$T : \mathbf{C} \to \mathbf{C}$$
one can ask the following:

Question: When is $T = U \circ F$ for some adjoint functors $F \dashv U$ to and from another category $\mathbf{D}$?

Thus, we seek necessary and sufficient conditions on the given endofunctor $T :$ $\mathbf{C} \to \mathbf{C}$ for recovering a category $\mathbf{D}$ and adjunction $F \dashv U$. Of course, not every T arises so, and we see that even if $T = U \circ F$ for *some* $\mathbf{D}$ and $F \dashv U$, we cannot always recover *that* adjunction. Thus, a better way to ask the question would be, given an adjunction what sort of "trace" does it leave on a category and can we recover the adjunction from this?

First, suppose we have $\mathbf{D}$ and $F \dashv U$ and T is the composite functor $T = U \circ F$. We then have a natural transformation,
$$\eta : 1 \to T.$$

And from the counit ϵ at FC,

$$\epsilon_{FC} : FUFC \to FC$$

we have $U\epsilon_{FC} : UFUFC \to UFC$, which we call,

$$\mu : T^2 \to T.$$

In general, then, as a first step toward answering our question, if T arises from an adjunction, then it should have such a structure $\eta : 1 \to T$ and $\mu : T^2 \to T$.

Now, what can be said about the structure (T, η, μ)? Actually, quite a bit! Indeed, the triangle equalities give us the following commutative diagrams:

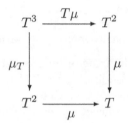

$$\mu \circ \mu T = \mu \circ T\mu \qquad\qquad (10.3)$$

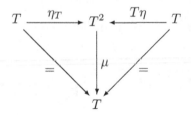

$$\mu \circ \eta T = 1_T = \mu \circ T\eta \qquad\qquad (10.4)$$

To prove the first one, for any $f : X \to Y$ in $\mathbf{D}$, the following square in $\mathbf{C}$ commutes, just since ϵ is natural:

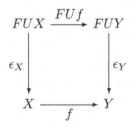

Now take $X = FUY$ and $f = \epsilon_Y$ to get the following:

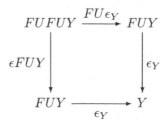

Putting FC for Y and applying U, therefore, gives this

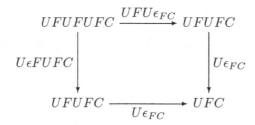

which has the required form (10.3). The equations (10.4) in the form

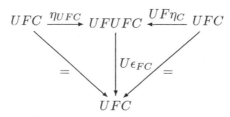

are simply the triangle identities, once taken at FC, and once under U. We record this data in the following definition.

Definition 10.2. A *monad* on a category $\mathbf{C}$ consists of an endofunctor $T : \mathbf{C} \to \mathbf{C}$, and natural transformations $\eta : 1_{\mathbf{C}} \to T$, and $\mu : T^2 \to T$ satisfying the two commutative diagrams above, that is,

$$\mu \circ \mu_T = \mu \circ T\mu \tag{10.5}$$

$$\mu \circ \eta_T = 1 = \mu \circ T\eta. \tag{10.6}$$

Note the formal analogy to the definition of a monoid. In fact, a monad is exactly the same thing as a *monoidal* monoid in the monoidal category $\mathbf{C}^{\mathbf{C}}$ with composition as the monoidal product, $G \otimes F = G \circ F$ (cf. section 7.8). For this reason, the equations (10.5) and (10.6) above are called the *associativity* and *unit* laws, respectively.

We have now shown the following proposition.

Proposition 10.3. *Every adjoint pair* $F \dashv U$ *with* $U : \mathbf{D} \to \mathbf{C}$, *unit* $\eta : UF \to 1_{\mathbf{C}}$ *and counit* $\epsilon : 1_{\mathbf{D}} \to FU$ *gives rise to a monad* (T, η, μ) *on* $\mathbf{C}$ *with*

$$T = U \circ F : \mathbf{C} \to \mathbf{C}$$

$$\eta : 1 \to T \quad \text{the unit}$$

$$\mu = U\epsilon_F : T^2 \to T.$$

Example 10.4. Let P be a poset. A monad on P is a monotone function $T : P \to P$ with $x \le Tx$ and $T^2 x \le Tx$. But then $T^2 = T$, that is, T is *idempotent*. Such a T, that is both inflationary and idempotent, is sometimes called a *closure operation* and written $Tp = \bar{p}$, since it acts like the closure operation on the subsets of a topological space. The "possibility operator" $\diamond p$ in modal logic is another example.

In the poset case, we can easily recover an adjunction from the monad. First, let $K = \mathrm{im}(T)(P)$ (the fixed points of T), and let $i : K \to P$ be the inclusion. Then let t be the factorization of T through K, as indicated in

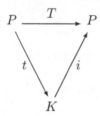

Observe that since $TTp = Tp$, for any element $k \in K$ we then have, for some $p \in P$, the equation $itik = ititp = itp = ik$, whence $tik = k$ since i is monic. We therefore have

$$p \le ik \quad \text{implies} \quad tp \le tik = k$$

$$tp \le k \quad \text{implies} \quad p \le itp \le ik$$

So indeed $t \dashv i$.

Example 10.5. Consider the covariant powerset functor

$$\mathcal{P} : \mathbf{Sets} \to \mathbf{Sets}$$

which takes each function $f : X \to Y$ to the image mapping $\mathrm{im}(f) : P(X) \to P(Y)$. Let $\eta_X : X \to \mathcal{P}(X)$ be the singleton operation

$$\eta_X(x) = \{x\}$$

and let $\mu_X : \mathcal{PP}(X) \to \mathcal{P}(X)$ be the union operation

$$\mu_X(\alpha) = \bigcup \alpha.$$

The reader should verify as an exercise that these operations are in fact natural in X and that this defines a monad $(\mathcal{P}, \{-\}, \bigcup)$ on **Sets**.

As we see in these examples, monads can, and often do, arise without coming from evident adjunctions. In fact, the notion of a monad originally did occur independently of adjunctions! Monads were originally also known by the names "triples" and sometimes "standard constructions." Despite their independent origin, however, our question "when does an endofunctor T arise from an adjunction?" has the simple answer: just if it is the functor part of a monad.

10.3 Algebras for a monad

Proposition 10.6. *Every monad arises from an adjunction. More precisely, given a monad (T, η, μ) on the category $\mathbf{C}$, there exists a category $\mathbf{D}$ and an adjunction $F \dashv U$, $\eta : 1 \to UF$, $\epsilon : FU \to 1$ with $U : \mathbf{D} \to \mathbf{C}$ such that*

$$T = U \circ F$$
$$\eta = \eta \quad \text{(the unit)}$$
$$\mu = U\epsilon_F.$$

Proof. We first define the important category $\mathbf{C}^T$ called the *Eilenberg–Moore category of T*. This will be our "$\mathbf{D}$." Then we need suitable functors

$$F : \mathbf{C} \rightleftarrows \mathbf{C}^T : U.$$

And, finally, we need natural transformations $\eta : 1 \to UF$ and $\epsilon : FU \to 1$ satisfying the triangle identities.

To begin, $\mathbf{C}^T$ has as *objects* the "T-algebras," which are pairs (A, α) of the form $\alpha : TA \to A$ in $\mathbf{C}$, such that

$$1_A = \alpha \circ \eta_A \quad \text{and} \quad \alpha \circ \mu_A = \alpha \circ T\alpha. \tag{10.7}$$

A *morphism* of T-algebras,

$$h : (A, \alpha) \to (B, \beta)$$

is simply an arrow $h : A \to B$ in $\mathbf{C}$, such that,

$$h \circ \alpha = \beta \circ T(h)$$

as indicated in the following diagram:

$$
\begin{array}{ccc}
TA & \xrightarrow{\;Th\;} & TB \\
\big\downarrow{\scriptstyle \alpha} & & \big\downarrow{\scriptstyle \beta} \\
A & \xrightarrow[h]{} & B
\end{array}
$$

It is obvious that $\mathbf{C}^T$ is a category with the expected composites and identities coming from $\mathbf{C}$, and that T is a functor.

Now define the functors,

$$U : \mathbf{C}^T \to \mathbf{C}$$

$$U(A, \alpha) = A$$

and

$$F : \mathbf{C} \to \mathbf{C}^T$$

$$FC = (TC, \mu_C).$$

We need to check that (TC, μ_C) is a T-algebra. The equations (10.7) for T-algebras in this case become

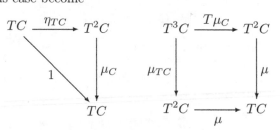

But these come directly from the definition of a monad.

To see that F is a functor, given any $h : C \to D$ in $\mathbf{C}$, we have

$$
\begin{array}{ccc}
T^2C & \xrightarrow{\;T^2h\;} & T^2D \\
\big\downarrow{\scriptstyle \mu_C} & & \big\downarrow{\scriptstyle \mu_D} \\
TC & \xrightarrow[Th]{} & TD
\end{array}
$$

since μ is natural. But this is a T-algebra homomorphism $FC \to FD$, so we can put

$$Fh = Th : TC \to TD$$

to get an arrow in $\mathbf{C}^T$.

Now we have defined the category $\mathbf{C}^T$ and the functors

$$\mathbf{C} \underset{U}{\overset{F}{\rightleftarrows}} \mathbf{C^T}$$

and we want to show that $F \dashv U$. Next, we need the unit and counit:

$$\bar{\eta} : 1_{\mathbf{C}} \to U \circ F$$
$$\epsilon : F \circ U \to 1_{\mathbf{C}^T}$$

Given $C \in \mathbf{C}$, we have

$$UF(C) = U(TC, \mu_C) = TC.$$

So we can take $\bar{\eta} = \eta : 1_{\mathbf{C}} \to U \circ F$, as required.

Given $(A, \alpha) \in \mathbf{C}^T$,

$$FU(A, \alpha) = (TA, \mu_A)$$

and the definition of a T-algebra makes the following diagram commute:

$$
\begin{array}{ccc}
T^2 A & \xrightarrow{\;T\alpha\;} & TA \\
\downarrow{\scriptstyle \mu_A} & & \downarrow{\scriptstyle \alpha} \\
TA & \xrightarrow[\;\alpha\;]{} & A
\end{array}
$$

But this is a morphism $\epsilon_{(A,\alpha)} : (TA, \mu_A) \to (A, \alpha)$ in $\mathbf{C}^T$. Thus we are setting

$$\epsilon_{(A,\alpha)} = \alpha.$$

And ϵ is *natural* by the definition of a morphism of T-algebras, as follows. Given any $h : (A, \alpha) \to (B, \beta)$, we need to show

$$h \circ \epsilon_{(A,\alpha)} = \epsilon_{(B,\beta)} \circ Th.$$

But by the definition of ϵ, that is, $h \circ \alpha = \beta \circ Th$, which holds since h is a T-algebra homomorphism.

Finally, the triangle identities now read as follows:

1. For (A, α), a T-algebra

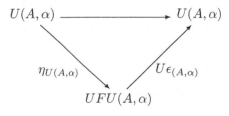

which amounts to

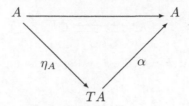

which holds since (A, α) is T-algebra.

2. For any $C \in \mathbf{C}$

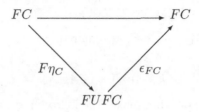

which is

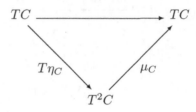

which holds by one of the unit laws for T.

Finally, note that we indeed have

$$T = U \circ F$$
$$\eta = \text{ unit of } F \dashv U.$$

And for the multiplication,

$$\bar{\mu} = U\epsilon F$$

we have, for any $C \in \mathbf{C}$,

$$\bar{\mu}_C = U\epsilon_{FC} = U\epsilon_{(TC, \mu_C)} = U\mu_C = \mu_C.$$

So $\bar{\mu} = \mu$ and we are done; the adjunction $F \dashv U$ via η and ϵ gives rise to the monad (T, η, μ). $\qquad\qquad\square$

Example 10.7. Take the free monoid adjunction,

$$F : \mathbf{Sets} \rightleftarrows \mathbf{Mon} : U.$$

The monad on **Sets** is then $T : \mathbf{Sets} \to \mathbf{Sets}$, where for any set X, $T(X) = UF(X) =$ "strings over X." The unit $\eta : X \to TX$ is the usual "string of length one" operation, but what is the multiplication?

$$\mu : T^2 X \to TX$$

Here $T^2 X$ is the set of strings of strings,

$$[[x_{11}, \ldots, x_{1n}], [x_{21}, \ldots, x_{2n}], \ldots, [x_{m1}, \ldots, x_{mn}]].$$

And μ of such a string of strings is the string of their elements,

$$\mu([[x_{11}, \ldots, x_{1n}], [x_{21}, \ldots, x_{2n}], \ldots, [x_{m1}, \ldots, x_{mn}]]) = [x_{11}, \ldots, x_{mn}].$$

Now, what is a T-algebra in this case? By the equations for a T-algebra, it is a map,

$$\alpha : TA \to A$$

from strings over A to elements of A, such that

$$\alpha[a] = a$$

and

$$\alpha(\mu([[\ldots], [\ldots], \ldots, [\ldots]])) = \alpha(\alpha[\ldots], \alpha[\ldots], \ldots, \alpha[\ldots]).$$

If we start with a monoid, then we can get a T-algebra $\alpha : TM \to M$ by

$$\alpha[m_1, \ldots, m_n] = m_1 \cdot \ldots \cdot m_n.$$

This clearly satisfies the required conditions. Observe that we can even recover the monoid structure from m by $u = m(-)$ for the unit and $x \cdot y = m(x, y)$ for the multiplication. Indeed, *every* T-algebra is of this form for a *unique* monoid (exercise!).

We have now given constructions back and forth between adjunctions and monads. And we know that if we start with a monad $T : \mathbf{C} \to \mathbf{C}$, and then take the adjunction,

$$F^T : \mathbf{C} \rightleftarrows \mathbf{C}^T : U^T$$

then we can get the monad back by $T = U^T \circ F^T$. Thus, in particular, every monoid arises from *some* adjunction. But are $\mathbf{C}^T, U^T, F^T$ unique with this property?

In general, the answer is *no*. There may be many different categories $\mathbf{D}$ and adjunctions $F \dashv U : \mathbf{D} \to \mathbf{C}$, all giving the same monad on $\mathbf{C}$. We have used the Eilenberg–Moore category $\mathbf{C}^T$, but there is also something called the "Kleisli category," which is in general different from $\mathbf{C}^T$, but also has an adjoint pair to $\mathbf{C}$ giving rise to the same monad (see the exercises).

If we start with an adjunction $F \dashv U$ and construct $\mathbf{C}^T$ for $T = U \circ F$, we then get a comparison functor $\Phi : \mathbf{D} \to \mathbf{C}^T$, with

$$U^T \circ \Phi \cong U$$

$$\Phi \circ F = F^T$$

In fact, Φ is *unique* with this property. A functor $U : \mathbf{D} \to \mathbf{C}$ is called *monadic* if it has a left adjoint $F \dashv U$, such that this comparison functor is an equivalence of categories,

$$\mathbf{D} \xrightarrow[\cong]{\Phi} \mathbf{C}^T$$

for $T = UF$.

Typical examples of monadic forgetful functors $U : \mathbf{C} \to \mathbf{Sets}$ are those from the "algebraic" categories arising as models for equational theories, like monoids, groups, rings, etc. Indeed, one can reasonably take monadicity as the *definition* of being "algebraic."

An example of a right adjoint that is *not* monadic is the forgetful functor from posets,

$$U : \mathbf{Pos} \to \mathbf{Sets}.$$

Its left adjoint F is the discrete poset functor. For any set X, therefore, one has as the unit the identity function $X = UF(X)$. The reader can easily show that the Eilenberg–Moore category for $T = 1_{\mathbf{Sets}}$ is then just $\mathbf{Sets}$ itself.

10.4 Comonads and coalgebras

By definition, a *comonad* on a category $\mathbf{C}$ is a monad on $\mathbf{C}^{\text{op}}$. Explicitly, this consists of an endofunctor $G : \mathbf{C} \to \mathbf{C}$ and natural transformations,

$$\epsilon : G \to 1 \qquad \text{the counit}$$

$$\delta : G \to G^2 \quad \text{comultiplication}$$

satisfying the duals of the equations for a monad, namely

$$\delta_G \circ \delta = G\delta \circ \delta$$

$$\epsilon_G \circ \delta = 1_G = G\epsilon \circ \delta.$$

We leave it as an exercise in duality to verify that an adjoint pair $F \dashv U$ with $U : \mathbf{D} \to \mathbf{C}$ and $F : \mathbf{C} \to \mathbf{D}$ and $\eta : 1_{\mathbf{C}} \to UF$ and $\epsilon : FU \to 1_{\mathbf{D}}$ gives rise to a comonad (G, ϵ, δ) on $\mathbf{D}$, where

$$G = F \circ U : \mathbf{D} \to \mathbf{D}$$

$$\epsilon : G \to 1$$

$$\delta = F\eta_U : G \to G^2.$$

The notions of coalgebra for a comonad, and of a comonadic functor, are of course also precisely dual to the corresponding ones for monads. Why do we even bother to study these notions separately, rather than just considering their duals? As in other examples of duality, there are actually two distinct reasons:

1. We may be interested in a particular category with special properties not had by its dual. A comonad on $\mathbf{Sets}^{\mathbf{C}}$ is of course a monad on $(\mathbf{Sets}^{\mathbf{C}})^{\mathrm{op}}$, but as we now know, $\mathbf{Sets}^{\mathbf{C}}$ has many special properties that its dual does not have (e.g., it is a topos!). So we can profitably consider the notion of a comonad on such a category.

   A simple example of this kind is the comonad $G = \Delta \circ \varprojlim$ resulting from composing the "constant functor" functor $\Delta : \mathbf{Sets} \to \mathbf{Sets}^{\mathbf{C}}$ with the "limit" functor $\varprojlim : \mathbf{Sets}^{\mathbf{C}} \to \mathbf{Sets}$. It can be shown in general that the coalgebras for this comonad again form a topos. In fact, they are just the constant functors $\Delta(S)$ for sets S, and the category $\mathbf{Sets}$ is thus comonadic over $\mathbf{Sets}^{\mathbf{C}}$.

2. It may happen that both structures—monad and comonad—occur together, and interact. Taking the opposite category will not alter this situation! This happens for instance when a system of *three* adjoint functors are composed:

$$L \dashv U \dashv R \qquad \mathbf{C} \; \underset{\underset{L}{\longleftarrow}}{\overset{\overset{R}{\longrightarrow}}{\xleftarrow{\;\;U\;\;}}} \; \mathbf{D}$$

resulting in a monad $T = U \circ L$ and a comonad $G = U \circ R$, both on $\mathbf{C}$. In such a case, T and G are then of course also adjoint $T \dashv G$.

   This arises, for instance, in the foregoing example with $R = \varprojlim$, and $U = \Delta$, and $L = \varinjlim$ the "colimit" functor. It also occurs in propositional modal logic, with $T = \Diamond$ "possibility" and $G = \Box$ "necessity," where the adjointness $\Diamond \dashv \Box$ is equivalent to the law known to modal logicians as "S5."

   A related example is given by the open and closed subsets of a topological space: the topological interior operation on arbitrary subsets is a comonad and closure is a monad. We leave the details as an exercise.

10.5 Algebras for endofunctors

Some very basic kinds of algebraic structures have a more simple description than as algebras for a monad, and this description generalizes to structures that are not algebras for any monad, but still have some algebra-like properties.

As a familiar example, consider first the underlying structure of the notion of a group. We have a set G equipped with operations as indicated in the following:

$$G \times G \xrightarrow{\;m\;} G \xleftarrow{\;i\;} G$$

$$u \Big\uparrow$$

$$1$$

We do not assume, however, that these operations satisfy the group equations of associativity, etc. Observe that this description of what we call a "group structure" can plainly be compressed into a single arrow of the form

$$1 + G + G \times G \xrightarrow{\;[u,i,m]\;} G$$

Now let us define the functor $F : \mathbf{Sets} \to \mathbf{Sets}$ by

$$F(X) = 1 + X + X \times X$$

Then a group structure is simply an arrow,

$$\gamma : F(G) \to G.$$

Moreover, a homomorphism of group structures in the conventional sense

$$h : G \to H,$$

$$h(u_G) = u_H$$
$$h(i(x)) = i(h(x))$$
$$h(m(x,y)) = m(h(x), h(y))$$

is then exactly a function $h : G \to H$ such that the following diagram commutes:

$$
\begin{array}{ccc}
F(G) & \xrightarrow{\;F(h)\;} & F(H) \\
\gamma \downarrow & & \downarrow \vartheta \\
G & \xrightarrow[\;h\;]{} & H
\end{array}
$$

where $\vartheta : F(H) \to H$ is the group structure on H. This observation motivates the following definition.

Definition 10.8. Given an endofunctor $P : \mathcal{S} \to \mathcal{S}$ on any category $\mathcal{S}$, a *P-algebra* consists of an object A of $\mathcal{S}$ and an arrow,

$$\alpha : PA \to A.$$

A *homomorphism* $h : (A, \alpha) \to (B, \beta)$ of P-algebras is an arrow $h : A \to B$ in $\mathcal{S}$ such that $h \circ \alpha = \beta \circ P(h)$, as indicated in the following diagram:

$$
\begin{array}{ccc}
P(A) & \xrightarrow{P(h)} & P(B) \\
\alpha \downarrow & & \downarrow \beta \\
A & \xrightarrow{\quad h \quad} & B
\end{array}
$$

The category of all such P-algebras and their homomorphisms are denoted as

$$P\text{-Alg}(\mathcal{S})$$

We usually write more simply P-Alg when $\mathcal{S}$ is understood. Also, if there is a monad present, we need to be careful to distinguish between algebras for the monad and algebras for the endofunctor (especially if P is the functor part of the monad!).

Example 10.9. 1. For the functor $P(X) = 1 + X + X \times X$ on **Sets**, we have already seen that the category **GrpStr** of group structures is the same thing as the category of P-algebras,

$$P\text{-Alg} = \textbf{GrpStr}.$$

2. Clearly, for any other algebraic structure of finite "signature," that is, consisting of finitely many, finitary operations, there is an analogous description of the structures of that sort as algebras for an associated endofunctor. For instance, a *ring structure*, with two nullary, one unary, and two binary operations is given by the endofunctor

$$R(X) = 2 + X + 2 \times X^2.$$

In general, a functor of the form

$$P(X) = C_0 + C_1 \times X + C_2 \times X^2 + \cdots + C_n \times X^n$$

with natural number coefficients C_k, is called a (finitary) *polynomial functor*, for obvious reasons. These functors present exactly the *finitary structures*. The same thing holds for finitary structures in any category $\mathcal{S}$

with finite products and coproducts; these can always be represented as algebras for a suitable endofunctor.

3. In a category such as **Sets** that is complete and cocomplete, there is an evident generalization to infinitary signatures by using generalized or "infinitary" polynomial functors, that is, ones with infinite sets C_k as coefficients (representing infinitely many operations of a given arity), infinite sets B_k as the exponents X^{B_k} (representing operations of infinite arity), or infinitely many terms (representing infinitely many different arities of operations), or some combination of these. The algebras for such an endofunctor

$$P(X) = \sum_{i \in I} C_i \times X^{B_i}$$

can then be naturally viewed as generalized "algebraic structures." Using locally cartesian closed categories, one can even present this notion without needing (co)completeness.

4. One can of course also consider algebras for an endofunctor $P : \mathcal{S} \to \mathcal{S}$ that is not polynomial at all, such as the covariant powerset functor $\mathcal{P} : \textbf{Sets} \to \textbf{Sets}$. This leads to a proper generalization of the notion of an "algebra," which however still shares some of the formal properties of conventional algebras, as seen below.

Let $P : \textbf{Sets} \to \textbf{Sets}$ be a polynomial functor, say

$$P(X) = 1 + X^2$$

(what structure is this?). Then the notion of an *initial* P-algebra gives rise to a recursion property analogous to that of the natural numbers. Specifically, let

$$[o, m] : 1 + I^2 \to I$$

be an initial P-algebra, that is, an initial object in the category of P-algebras. Then, explicitly, we have the structure

$$o \in I, \qquad m : I \times I \to I$$

and for any set X with a distinguished element and a binary operation

$$a \in X, \qquad * : X \times X \to X$$

there is a unique function $u : I \to X$ such that the following diagram commutes:

$$\begin{array}{ccc} 1 + I^2 & \xrightarrow{P(u)} & 1 + X^2 \\ {\scriptstyle [o,m]}\downarrow & & \downarrow{\scriptstyle [a,*]} \\ I & \xrightarrow[u]{} & X \end{array}$$

This of course says that, for all $i, j \in I$,

$$u(o) = a$$
$$u(m(i,j)) = u(i) * u(j)$$

which is exactly a *definition by structural recursion* of the function $u : I \to X$. Indeed, the usual recursion property of the natural numbers $\mathbb{N}$ with zero $0 \in \mathbb{N}$ and successor $s : \mathbb{N} \to \mathbb{N}$ says precisely that $(\mathbb{N}, 0, s)$ is the initial algebra for the endofunctor,

$$P(X) = 1 + X : \mathbf{Sets} \to \mathbf{Sets}$$

as the reader should check.

We next briefly investigate the question: When does an endofunctor have an initial algebra? The existence is constrained by the fact that initial algebras, when they exist, must have the following noteworthy property.

Lemma 10.10 (Lambek). *Given any endofunctor* $P : \mathcal{S} \to \mathcal{S}$ *on an arbitrary category* $\mathcal{S}$, *if* $i : P(I) \to I$ *is an initial P-algebra, then* i *is an isomorphism,*

$$P(I) \cong I.$$

We leave the proof as an easy exercise.

In this sense, the initial algebra for an endofunctor $P : \mathcal{S} \to \mathcal{S}$ is a "least fixed point" for P. Such algebras are often used in computer science to model "recursive datatypes" determined by the so-called fixed point equations $X = P(X)$.

Example 10.11. 1. For the polynomial functor,

$$P(X) = 1 + X^2$$

(monoid structure!), let us "unwind" the initial algebra,

$$[*, @] : 1 + I \times I \cong I.$$

Given any element $x \in I$, it is thus either of the form $*$ or of the form $x_1 @ x_2$ for some elements $x_1, x_2 \in I$. Each of these x_i, in turn, is either of the form $*$ or of the form $x_{i1} @ x_{i2}$, and so on. Continuing in this way, we have a representation of x as a finite, binary tree. For instance, an element

of the form $x = *@(*@*)$ looks like

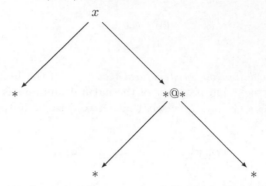

We can present the monoid structure explicitly by letting

$$I = \{t \mid t \text{ is a finite, binary tree}\}$$

with

$$* = \text{"the empty tree"}$$

$$@(t_1, t_2) = t_1 @ t_2$$

$$=$$

$$t_1 @ t_2$$

$$t_1 \qquad\qquad t_2$$

The isomorphism,

$$[*, @] : 1 + I \times I \to I$$

here is plain to see.

2. Similarly, for any other polynomial functor,

$$P(X) = C_0 + C_1 \times X + C_2 \times X^2 + \cdots + C_n \times X^n$$

we can describe the initial algebra (in **Sets**),

$$P(I) \cong I$$

as a set of trees with branching types and labels determined by P. For instance, consider the polynomial

$$P(X) = 1 + A \times X$$

for some set A. What is the initial algebra? Since,

$$[*, @] : 1 + A \times I \cong I$$

we can unwind an element x as

$$x = * \text{ or } a_1@x_1$$

$$x_1 = * \text{ or } a_2@x_2$$

$$\cdots$$

Thus, we essentially have $x = a_1@a_2@\cdots@a_n$. So I can be represented as the set A-List of (finite) lists of elements $a_1, a_2, \ldots$ of A, with the structure

$$* = \text{"the empty list"}$$

$$@(a, \ell) = a@\ell$$

The usual procedure of "recursive definition" follows from initiality. For example, the length function for lists length : A-List $\to \mathbb{N}$ is usually defined by

$$\text{length}(*) = 0 \tag{10.8}$$

$$\text{length}(a@\ell) = 1 + \text{length}(\ell) \tag{10.9}$$

We can do this by equipping $\mathbb{N}$ with a suitable $P(X) = 1 + A \times X$ structure, namely,

$$[0, m] : 1 + A \times \mathbb{N} \to \mathbb{N}$$

where $m(a, n) = 1+n$ for all $n \in \mathbb{N}$. Then by the universal mapping property of the initial algebra, we get a unique function length : A-List $\to \mathbb{N}$ making a commutative square:

$$
\begin{array}{ccc}
1 + A \times A\text{-List} & \xrightarrow{1 + A \times \text{length}} & 1 + A \times \mathbb{N} \\
\downarrow{\scriptstyle [*, @]} & & \downarrow{\scriptstyle [0, m]} \\
A\text{-List} & \xrightarrow[\text{length}]{} & \mathbb{N}
\end{array}
$$

But this commutativity is, of course, precisely equivalent to the equations (10.8) and (10.9).

In virtue of Lambek's lemma, we at least know that not all endofunctors can have initial algebras. For, consider the covariant powerset functor $\mathcal{P}$: **Sets** $\to$ **Sets**. An initial algebra for this would give us a set I with the property that $\mathcal{P}(I) \cong I$, which is impossible by the well-known theorem of Cantor! The following proposition gives a useful sufficient condition for the existence of an initial algebra.

Proposition 10.12. *If the category $\mathcal{S}$ has an initial object 0 and colimits of diagrams of type ω (call them "ω-colimits"), and the functor*

$$P : \mathcal{S} \to \mathcal{S}$$

preserves ω-colimits, then P has an initial algebra.

Proof. Note that this generalizes a very similar result for posets already given above as proposition 5.34. And even the proof by "Newton's method" is essentially the same! Take the ω-sequence

$$0 \to P0 \to P^2 0 \to \cdots$$

and let I be the colimit

$$I = \varinjlim_n P^n 0.$$

Then, since P preserves the colimit, there is an isomorphism

$$P(I) = P(\varinjlim_n P^n 0) \cong \varinjlim_n P(P^n 0) = \varinjlim_n P^n 0 = I$$

which is seen to be an initial algebra for P by an easy diagram chase. □

Since (as the reader should verify) every polynomial functor $P : \mathbf{Sets} \to \mathbf{Sets}$ preserves ω-colimits, we have

Corollary 10.13. *Every polynomial functor $P : \mathbf{Sets} \to \mathbf{Sets}$ has an initial algebra.*

Finally, we ask, what is the relationship between algebras for endofunctors and algebras for monads? The following proposition, which is a sort of "folk theorem," gives the answer.

Proposition 10.14. *Let the category $\mathcal{S}$ have finite coproducts. Given an endofunctor $P : \mathcal{S} \to \mathcal{S}$, the following conditions are equivalent:*

1. *The P-algebras are the algebras for a monad. Precisely, there is a monad $(T : \mathcal{S} \to \mathcal{S}, \eta, \mu)$, and an equivalence*

$$P\text{-Alg}(\mathcal{S}) \simeq \mathcal{S}^T$$

   between the category of P-algebras and the category $\mathcal{S}^T$ of algebras for the monad. Moreover, this equivalence preserves the respective forgetful functors to $\mathcal{S}$.

2. *The forgetful functor $U : P\text{-Alg}(\mathcal{S}) \to \mathcal{S}$ has a left adjoint*

$$F \vdash U.$$

3. *For each object A of $\mathcal{S}$, the endofunctor*

$$P_A(X) = A + P(X) : \mathcal{S} \to \mathcal{S}$$

   has an initial algebra.

Proof. That (1) implies (2) is clear.

For (2) implies (3), suppose that U has a left adjoint $F : \mathcal{S} \to P\text{-Alg}$ and consider the endofunctor $P_A(X) = A + P(X)$. An algebra (X, γ) is a map $\gamma : A + P(X) \to X$. But there is clearly a unique correspondence between the following three types of things:

$$\gamma : A + P(X) \to X$$

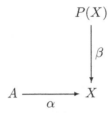

$$\alpha : A \to U(X, \beta)$$

Thus, the P_A-algebras can be described equivalently as arrows of the form $\alpha : A \to U(X, \beta)$ for P-algebras (X, β). Moreover, a P_A-homomorphism $h : (\alpha, U(X, \beta)) \to (\alpha', U(X', \beta'))$ is just a P-homomorphism $h : (X, \beta) \to (X', \beta')$ making a commutative triangle with α and $\alpha' : A \to U(X', \beta')$. But an initial object in this category is given by the unit $\eta : A \to UFA$ of the adjunction $F \vdash U$, which shows (3).

Indeed, given just the forgetful functor $U : P\text{-Alg} \to \mathcal{S}$, the existence of initial objects in the respective categories of arrows $\alpha : A \to U(X, \beta)$, for each A, is exactly what is needed for the existence of a left adjoint F to U. So (3) also implies (2).

Before concluding the proof, it is illuminating to see how the free functor $F : \mathcal{S} \to P\text{-Alg}$ results from condition (3). For each object A in $\mathcal{S}$, consider the initial P_A-algebra $\alpha : A + P(I_A) \to I_A$. In the notation of recursive type theory,

$$I_A = \mu_X.A + P(X)$$

meaning it is the (least) solution to the "fixed point equation"

$$X = A + P(X).$$

Since α is a map on the coproduct $A + P(I_A)$, we have $\alpha = [\alpha_1, \alpha_2]$, and we can let

$$F(A) = (I_A, \ \alpha_2 : P(I_A) \to I_A)$$

To define the action of F on an arrow $f : A \to B$, let $\beta : B + P(I_B) \to I_B$ be the initial P_B-algebra and consider the diagram

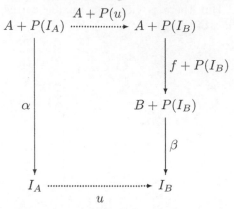

The right-hand vertical composite $\beta \circ (f + P(I_B))$ now makes I_B into a P_A-algebra. There is thus a unique P_A-homomorphism u as indicated, and we can set

$$F(f) = u.$$

Finally, to conclude, the fact that (2) implies (1) is an easy application of Beck's Precise Tripleability Theorem, for which we refer the reader to section VI.7 of Mac Lane's *Categories Work* (1971). □

10.6 Exercises

1. Let $\mathbb{T}$ be the equational theory with one constant symbol and one unary function symbol (no axioms). In any category with a terminal object, a natural numbers object (NNO) is just an initial $\mathbb{T}$-model. Show that the natural numbers

$$(\mathbb{N}, 0 \in \mathbb{N}, n + 1 : \mathbb{N} \to \mathbb{N})$$

is an NNO in **Sets**, and that any NNO is uniquely isomorphic to it (as a $\mathbb{T}$-model).

Finally, show that $(\mathbb{N}, 0 \in \mathbb{N}, n + 1 : \mathbb{N} \to \mathbb{N})$ is uniquely characterized (up to isomorphism) as the initial algebra for the endofunctor $F(X) = X + 1$.

2. Let $\mathbf{C}$ be a category and $T : \mathbf{C} \to \mathbf{C}$ an endofunctor. A T-*algebra* consists of an object A and an arrow $a : TA \to A$ in $\mathbf{C}$. A morphism $h : (a, A) \to (b, B)$ of T-algebras is a $\mathbf{C}$-morphism $h : A \to B$ such that $h \circ a = b \circ T(h)$. Let $\mathbf{C}$ be a category with a terminal object 1 and binary coproducts. Let $T : \mathbf{C} \to \mathbf{C}$ be the evident functor with object-part $C \mapsto C + 1$ for all objects C of $\mathbf{C}$. Show (easily) that the categories of T-algebras and $\mathbb{T}$-models ($\mathbb{T}$ as above) (in $\mathbf{C}$) are equivalent:

$$T\text{-Alg} \simeq \mathbb{T}\text{-Mod}.$$

Conclude that free T-algebras exist in **Sets**, and that an initial T-algebra is the same thing as an NNO.

3. ("Lambek's lemma") Show that for any endofunctor $T : \mathbf{C} \to \mathbf{C}$, if $i : TI \to I$ is an initial T-algebra, then i is an isomorphism. (Hint: consider a diagram of the following form, with suitable arrows.)

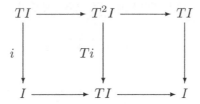

Conclude that for any NNO N in any category, there is an isomorphism $N + 1 \cong N$. Also, derive the usual recursion property of the natural numbers from initiality.

4. Given categories $\mathbf{C}$ and $\mathbf{D}$ and adjoint functors $F : \mathbf{C} \to \mathbf{D}$ and $U : \mathbf{D} \to \mathbf{C}$ with $F \dashv U$, unit $\eta : 1_{\mathbf{C}} \to UF$, and counit $\epsilon : FU \to 1_{\mathbf{D}}$, show that

$$T = U \circ F : \mathbf{C} \to \mathbf{C}$$

$$\eta : 1_{\mathbf{C}} \to T$$

$$\mu = U\epsilon_F : T^2 \to T$$

do indeed determine a monad on $\mathbf{C}$, as stated in the text.

5. Assume given categories $\mathbf{C}$ and $\mathbf{D}$ and adjoint functors

$$F : \mathbf{C} \rightleftarrows \mathbf{D} : U$$

with unit $\eta : 1_{\mathbf{C}} \to UF$ and counit $\epsilon : FU \to 1_{\mathbf{D}}$. Show that every D in $\mathbf{D}$ determines a $T = UF$ algebra $U\epsilon : UFUD \to UD$, and that there is a "comparison functor" $\Phi : \mathbf{D} \to \mathbf{C}^T$ which, moreover, commutes with the "forgetful" functors $U : \mathbf{D} \to \mathbf{C}$ and $U^T : \mathbf{C}^T \to \mathbf{C}$.

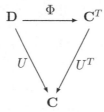

6. Show that $(P, s, \cup)$ is a monad on **Sets**, where

- $P : \mathbf{Sets} \to \mathbf{Sets}$ is the covariant powerset functor, which takes each function $f : X \to Y$ to the image mapping

$$P(f) = im(f) : P(X) \to P(Y)$$

- for each set X, the component $s_X : X \to P(X)$ is the singleton mapping, with

$$s_X(x) = \{x\} \subseteq X$$

for each $x \in X$;

- for each set X, the component $\cup_X : PP(X) \to P(X)$ is the union operation, with

$$\cup_X(\alpha) = \{x \in X \mid \exists_{U \in \alpha}.\ x \in U\} \subseteq X$$

for each $\alpha \subseteq P(X)$.

7. Determine the category of (Eilenberg–Moore) algebras for the $(P, s, \cup)$ monad on **Sets** defined in the foregoing problem. (Hint: consider complete lattices.)

8. Consider the free $\dashv$ forgetful adjunction

$$F : \mathbf{Sets} \rightleftarrows \mathbf{Mon} : U$$

between sets and monoids, and let (T, η^T, μ^T) be the associated monad on **Sets**. Show that any T-algebra $\alpha : TA \to A$ for this monad comes from a monoid structure on A (exhibit the monoid multiplication and unit element).

9. (a) Show that an adjoint pair $F \dashv U$ with $U : \mathbf{D} \to \mathbf{C}$ and $\eta : UF \to 1_\mathbf{C}$ and $\epsilon : 1_\mathbf{D} \to FU$ also gives rise to a *comonad* (G, ϵ, δ) in $\mathbf{D}$, with

$$G = F \circ U : \mathbf{D} \to \mathbf{D}$$

$$\epsilon : G \to 1 \text{ the counit}$$

$$\delta = F\eta_U : G \to G^2$$

satisfying the duals of the equations for a monad.

(b) Define the notion of a *coalgebra* for a comonad, and show (by duality) that every comonad (G, ϵ, δ) on a category $\mathbf{D}$ "comes from" a (not necessarily unique) adjunction $F \dashv G$ such that $G = FU$ and ϵ is the counit.

(c) Let **End** be the category of sets equipped with an endomorphism, $e : S \to S$. Consider the functor $G : \mathbf{End} \to \mathbf{End}$ defined by

$$G(S, e) = \{x \in S \mid e^{(n+1)}(x) = e^{(n)}(x) \text{ for some } n\}$$

equipped with the restriction of e. Show that this is the functor part of a comonad on **End**.

10. Verify that the open and closed subsets of a topological space give rise to comonad and monad, respectively, on the powerset of the underlying pointset. Moreover, the categories of coalgebras and algebras are isomorphic.

11. (Kleisli category) Given a monad (T, η, μ) on a category $\mathbf{C}$, in addition to the Eilenberg–Moore category, we can construct another category $\mathbf{C}_T$ and an adjunction $F \dashv U$, $\eta : 1 \to UF$, $\epsilon : FU \to 1$ with $U : \mathbf{C}_T \to \mathbf{C}$ such that

$$T = U \circ F$$

$$\eta = \eta \quad \text{(the unit)}$$

$$\mu = U\epsilon_F$$

This category $\mathbf{C}_T$ is called the *Kleisli category* of the adjunction, and is defined as follows:

- the objects are the same as those of $\mathbf{C}$, but written $A_T, B_T, \ldots$,
- an arrow $f_T : A_T \to B_T$ is an arrow $f : A \to TB$ in $\mathbf{C}$,
- the identity arrow $1_{A_T} : A_T \to A_T$ is the arrow $\eta_A : A \to TA$ in $\mathbf{C}$,
- for composition, given $f_T : A_T \to B_T$ and $g_T : B_T \to C_T$, the composite $g_T \circ f_T : A_T \to C_T$ is defined to be

$$\mu_C \circ Tg_T \circ f_T$$

as indicated in the following diagram:

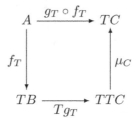

Verify that this indeed defines a category, and that there are adjoint functors $F : \mathbf{C} \to \mathbf{C}_T$ and $U : \mathbf{C}_T \to \mathbf{C}$ giving rise to the monad as $T = UF$, as claimed.

12. Let $P : \mathbf{Sets} \to \mathbf{Sets}$ be a polynomial functor,

$$P(X) = C_0 + C_1 \times X + C_2 \times X^2 + \cdots + C_n \times X^n$$

with natural number coefficients C_k. Show that P preserves ω-colimits.

13. The notion of a *coalgebra* for an endofunctor $P : \mathcal{S} \to \mathcal{S}$ on an arbitrary category $\mathcal{S}$ is exactly dual to that of a P-algebra. Determine the *final* coalgebra for the functor

$$P(X) = 1 + A \times X$$

for a set A. (Hint: Recall that the initial algebra consisted of *finite* lists $a_1, a_2, \ldots$ of elements of A.)

SOLUTIONS TO SELECTED EXERCISES

Chapter 1

1. (a) Identity arrows behave correctly, for if $f \subset A \times B$, then

$$f \circ 1_A = \{\langle a, b \rangle \mid \exists a' \in A : \langle a, a' \rangle \in 1_A \wedge \langle a', b \rangle \in f\}$$
$$= \{\langle a, b \rangle \mid \exists a' \in A : a = a' \wedge \langle a', b \rangle \in f\}$$
$$= \{\langle a, b \rangle \mid \langle a, b \rangle \in f\} = f$$

and symmetrically $1_B \circ f = f$. Composition is associative; if $f \subseteq A \times B$, $g \subseteq B \times C$, and $h \subseteq C \times D$, then

$$(h \circ g) \circ f = \{\langle a, d \rangle \mid \exists b : \langle a, b \rangle \in f \wedge \langle b, d \rangle \in h \circ g\}$$
$$= \{\langle a, d \rangle \mid \exists b : \langle a, b \rangle \in f \wedge \langle b, d \rangle \in \{\langle b, d \rangle \mid \exists c : \langle b, c \rangle \in g \wedge \langle c, d \rangle \in h\}\}$$
$$= \{\langle a, d \rangle \mid \exists b : \langle a, b \rangle \in f \wedge \exists c : \langle b, c \rangle \in g \wedge \langle c, d \rangle \in h\}$$
$$= \{\langle a, d \rangle \mid \exists b \exists c : \langle a, b \rangle \in f \wedge \langle b, c \rangle \in g \wedge \langle c, d \rangle \in h\}$$
$$= \{\langle a, d \rangle \mid \exists c : (\exists b : \langle a, b \rangle \in f \wedge \langle b, c \rangle \in g) \wedge \langle c, d \rangle \in h\}$$
$$= \{\langle a, d \rangle \mid \exists c : \langle a, b \rangle \in g \circ f \wedge \langle c, d \rangle \in h\}$$
$$= h \circ (g \circ f).$$

2. (a) **Rel** $\cong$ **Rel**^{op}. The isomorphism functor (in both directions) takes an object A to itself, and takes a relation $f \subseteq A \times B$ to the *opposite relation* $f^{\mathrm{op}} \subseteq B \times A$ defined by $f^{\mathrm{op}} := \{\langle b, a \rangle \mid \langle a, b \rangle \in f\}$. It is straightforward to check that this is a functor **Rel** $\to$ **Rel**^{op} and **Rel**$^{\mathrm{op}} \to$ **Rel**, and it is its own inverse.

   (b) **Sets** $\ncong$ **Sets**^{op}. Consider maps into the empty set $\emptyset$; there is exactly one. If **Sets** $\cong$ **Sets**^{op} held, there would have to be a corresponding set $\emptyset'$ with exactly one arrow out of it.

   (c) $P(X) \cong P(X)^{\mathrm{op}}$. The isomorphism takes each element U of the powerset to its complement $X - U$. Functoriality amounts to the fact that $U \subseteq V$ implies $X - V \subseteq X - U$.

3. (a) A bijection f from a set A to a set B, and its inverse f^{-1}, comprise an isomorphism; $f(f^{-1}(b)) = b$ and $f^{-1}(f(a)) = a$, and so $f \circ f^{-1} = 1_B$ and $f^{-1} \circ f = 1_A$, by definition of the inverse. If an arrow $f : A \to B$ in **Sets** is an isomorphism, then there is an arrow $g : B \to A$ such that $f \circ g = 1_B$ and $g \circ f = 1_A$. The arrow f is an injection because $f(a) = f(a')$ implies $a = g(f(a)) = g(f(a')) = a'$, and f is surjective because every $b \in B$ has a preimage, namely $g(b)$, since $f(g(b)) = b$.

(b) Monoid homomorphisms that are isomorphisms are also isomorphisms in **Sets**, so by the previous solution they are bijective homomorphisms. It remains to show that bijective homomorphisms are isomorphisms. It is sufficient to show that the inverse mapping of a bijective homomorphism $f : M \to N$ is a homomorphism. But we have

$$f^{-1}(b \star_N b') = f^{-1}(f(f^{-1}(b)) \star_N f(f^{-1}(b')))$$
$$= f^{-1}(f(f^{-1}(b) \star_M f^{-1}(b')))$$
$$= f^{-1}(b) \star_M f^{-1}(b')$$

and $f^{-1}(e_N) = f^{-1}(f(e_M)) = e_M$.

(c) Consider the posets $A = (U, \leq_A)$ and $B = (U, \leq_B)$ given by $U = \{0, 1\}$, $\leq_A = \{\langle 0, 0 \rangle, \langle 1, 1 \rangle\}$, and $\leq_B = \{\langle 0, 0 \rangle, \langle 0, 1 \rangle, \langle 1, 1 \rangle\}$. The identity function $i : U \to U$ is an arrow $A \to B$ in **Posets**, and it is a bijection, but the only arrows $B \to A$ in **Posets** are the two constant functions $U \to U$, because arrows in **Posets** must be monotone. Neither is an inverse to i, which is therefore not an isomorphism.

6. The *coslice category* $C/\mathbf{C}$, the category whose objects are arrows $f : C \to A$ for $A \in \mathbf{C}$ and whose arrows $f \to f'$ are arrows h completing commutative triangles

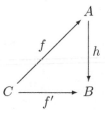

can equivalently be described as $(\mathbf{C}^{\mathrm{op}}/C)^{\mathrm{op}}$. For example, in the above diagram f, f' are arrows *into* C in the opposite category $\mathbf{C}^{\mathrm{op}}$, so they are objects in the slice $\mathbf{C}^{\mathrm{op}}/C$. The arrow h is $B \to A$ in $\mathbf{C}^{\mathrm{op}}$ and $h \circ f = f'$ so it is an arrow $B \to A$ in $\mathbf{C}^{\mathrm{op}}/C$, hence an arrow $A \to B$ in $(\mathbf{C}^{\mathrm{op}}/C)^{\mathrm{op}}$.

9. The free category on the graph

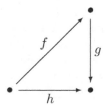

is

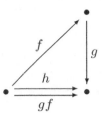

plus three identity arrows, one for each object. The free category on the graph

has infinitely many arrows, all possible finite sequence of alternating fs and gs—there are two empty sequences (i.e., identity arrows), one for each object.

10. The graphs whose free categories have exactly six arrows are the discrete graph with six nodes, and the following 10 graphs:

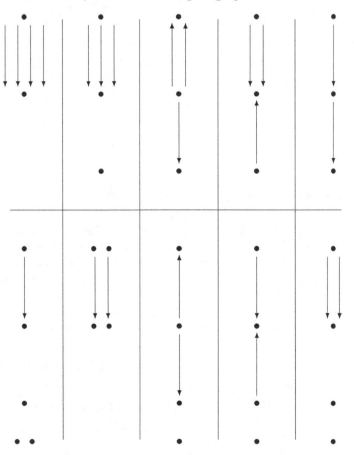

11. (a) The functor $M : \mathbf{Sets} \to \mathbf{Mon}$ that takes a set X to the free monoid on X (i.e., strings over X and concatenation) and takes a function $f : X \to Y$ to the function $M(f)$ defined by $M(f)(a_1 \ldots a_k) = f(a_1) \ldots f(a_k)$ is a functor; $M(f)$ is a monoid homomorphism $MX \to MY$ since it preserves the monoid identity (the empty string) and the monoid operation (composition). It can be checked that M preserves identity functions and composition: $M(1_X)(a_1 \ldots a_k) = 1_X(a_1) \ldots 1_X(a_k) = a_1 \ldots a_k$ and

$$M(g \circ f)(a_1 \ldots a_k) = (g \circ f)(a_1) \ldots (g \circ f)(a_k)$$
$$= g(f(a_1)) \ldots g(f(a_k)) = M(g)(M(f)(a_1 \ldots a_k))$$
$$= (M(g) \circ M(f))(a_1 \ldots a_k).$$

12. Let $\mathbf{D}$ be a category and $h : G \to U(\mathbf{D})$ be a graph homomorphism. Suppose $\bar{h}$ is a functor $\mathbf{C}(G) \to \mathbf{D}$ such that

$$U(\bar{h}) \circ i = h \qquad\qquad (*).$$

From this equation, we see that $U(\bar{h})(i(x)) = h(x)$ for all vertices and edges $x \in G$. So the behavior of $\bar{h}$ on objects and paths of length one (i.e., arrows in the image of i) in $\mathbf{C}(G)$ is completely determined by the requirement $(*)$. But since $\bar{h}$ is assumed to be a functor, and so must preserve composition, its behavior on arrows in $\mathbf{C}(G)$ that correspond to longer paths in G is also determined, by a simple induction. Now it must be that $\bar{h}(f_1 \cdots f_k) = h(f_1) \circ \cdots h(f_k)$ if $\bar{h}$ is a functor, and similarly $\bar{h}(\varepsilon_A) = 1_A$, where ε_A is the empty path at A. So uniqueness of $\bar{h}$ is established, and it is easily checked that this definition is indeed a functor, so the UMP is satisfied.

Chapter 2

1. Suppose $f : A \to B$ is epi and not surjective. Choose $b \in B$ not in the range of f. Define $g_1, g_2 : B \to \{0,1\}$ as follows: $g_1(x) = 0$ for all $x \in B$, and $g_2(x) = 1$ if $x = b$, and 0 otherwise. Note that $g_1 \circ f = g_2 \circ f$ by choice of b, a contradiction. In the other direction, suppose f is surjective, and suppose $g_1, g_2 : B \to C$ are such that $g_1 \neq g_2$. Then there is $b \in B$ such that $g_1(b) \neq g_2(b)$. By assumption, b has a preimage a such that $f(a) = b$. So $g_1(f(a)) \neq g_2(f(a))$ and $g_1 \circ f \neq g_2 \circ f$.

4. (a) Iso: the inverse of h is $f^{-1} \circ g^{-1}$. Monic: If $h \circ k_1 = h \circ k_2$, then $g \circ f \circ k_1 = g \circ f \circ k_2$. Since g is monic, $f \circ k_1 = f \circ k_2$. Since f is monic, $k_1 = k_2$. Epi: dual argument.

   (b) If $f \circ k_1 = f \circ k_2$, then $g \circ f \circ k_1 = g \circ f \circ k_2$. Since h is monic, $k_1 = k_2$.

   (c) Dual argument to (b).

   (d) In **Sets**, put $A = C = \{0\}$, $B = \{0,1\}$, and all arrows constantly 0. h is monic but g is not.

5. Suppose $f : A \to B$ is an isomorphism. Then f is mono because $f \circ k_1 = f \circ k_2$ implies $k_1 = f^{-1} \circ f \circ k_1 = f^{-1} \circ f \circ k_2 = k_2$, and dually f is mono also. Trivially, f is split mono and split epi because $f \circ f^{-1} = 1_B$ and $f^{-1} \circ f = 1_A$. So we know $(a) \Rightarrow (b), (c), (d)$. If f is mono and split epi, then there is g such that $f \circ g = 1_B$. But since f is mono, $(f \circ g) \circ f = f \circ (g \circ f) = f = f \circ 1_A$ implies $g \circ f = 1_A$ and so g is in fact the inverse of f, and we have $(b) \Rightarrow (a)$. Dually, $(c) \Rightarrow (a)$. The fact that $(d) \Rightarrow (b), (c)$ needs only that split mono implies mono (or dually that split epi implies epi). If there is g such that $g \circ f = 1_A$, then $f \circ k_1 = f \circ k_2$ implies $k_1 = g \circ f \circ k_1 = g \circ f \circ k_2 = k_2$.

6. If $h : G \to H$ is injective on edges and vertices, and $h \circ f = h \circ g$ in **Graphs**, then the underlying set functions on edges and vertices are mono arrows in **Sets**, so the edge and vertex parts of f and g are equal, and so $f = g$. If $h : G \to H$ is mono in **Graphs**, and it is not injective on vertices, then there are two vertices v, w such that $h(v) = h(w)$. Let 1 be the graph with one vertex, and f, g be graph homomorphisms $1 \to G$ taking that vertex to v, w, respectively. Then, $h \circ f = h \circ g$. A similar argument holds for edges.

9. First, in the category **Pos**, an arrow is epi iff it is surjective: suppose that $f : A \to B$ is surjective and let $g, h : B \to C$ with $gf = hf$. In **Pos**, this means that g and h agree on the image of f, which by surjectivity is all of B. Hence $g = h$ and f is epi. On the other hand, suppose f is not epi and that $g, h : B \to C$ witness this. Since $g \neq h$, there is some $b \in B$ with $g(b) \neq h(b)$. But from this $b \notin f(A)$, and so A is not surjective.

Next, the singleton set **1**, regarded as a poset, is projective: suppose $f : 1 \to Y$ and $e : X \twoheadrightarrow Y$ are arrows in **Pos**, with e epi. Then e is surjective, so there is some $x \in X$ with $e(x) = f(*)$. Any map $* \mapsto x$ witnesses the projectivity of **1**.

10. Any set A is projective in **Pos**: suppose that $f : A \to Y$ and $e : X \twoheadrightarrow Y$ are arrows in **Pos**. Choose for each $y \in Y$ an element $x_y \in X$ with $f(x_y) = y$; this is possible since e is epi and hence surjective. Now define a map $\bar{f} : A \to X$ by $a \mapsto x_{f(a)}$. Since A is discrete this is necessarily monotonic, and we have $e\bar{f} = f$, so A is projective.

For contrast, the two element poset $P = \{0 \leq 1\}$ is not projective. Indeed, we may take f to be the identity and X to be the discrete two-element set $\{a, b\}$. Then the surjective map $e : a \mapsto 0, b \mapsto 1$ is an epi, since it is surjective. However, any monotone map $g : P \to \{a, b\}$ must identify 0 and 1, since the only arrows in the second category are identities. But then $e \circ g \neq 1_P$. Thus, there is no function g lifting the identity map on P across e, so P is not projective.

Moreover, every projective object in **Pos** is discrete: For suppose Q is projective. We can always consider the discretation $|Q|$ of Q, which has the same objects as Q and only identity arrows. We clearly get a map $|Q| \to Q$ which is surjective and hence epi. This means that we can complete the diagram

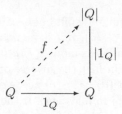

But the only object function that could possibly commute in this situation is the object identity. Then,

$$x \leq x' \iff f(x) \leq f(x') \iff f(x) = f(x') \iff x = x'.$$

But then the only arrows of Q are identity arrows, so Q is discrete, as claimed. Thus, the projective posets are exactly the discrete sets. Clearly, composition of maps and identity arrows of discrete posets are exactly those of **Set**, so **Set** is a subcategory of **Pos**. Moreover, every function between discrete sets is monotone, so this is a full subcategory.

11. The UMP of a free monoid states that for any $f : A \to UB$, there is a unique $\bar{f} : MA \to B$ such that

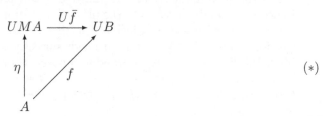

$$(*)$$

commutes. For $\eta : A \to UM$ to be an initial object in A-**Mon**, it must be that for object $f : A \to UB$, there is a unique arrow $\bar{f}$ in A-**Mon** from η to f. But the definition of arrow in A-**Mon** is such that this arrow must complete exactly the commutative triangle $(*)$ above. Therefore, the two characterizations of the free monoid coincide.

13. Let P be the iterated product $A \times (B \times C)$ with the obvious maps $p_1 : P \to A$, $p_2 : P \to B \times C \to B$, and $p_3 : P \to B \times C \to C$. Define $Q = (A \times B) \times C$ and q_i similarly. By the UMP, we get a unique map $f_1 = p_1 \times p_2 : P \to A \times B$. Applying it again, we get a unique map $f = (p_1 \times p_2) \times p_3 : P \to Q$ with $q_i f = p_i$. We can run a similar argument to get a map g in the other direction. Composing, we get $gf : P \to P$ which respects the p_i. By the UMP, such a map is unique, but the identity is another such map. Thus they must be the same, so $gf = 1_P$. Similarly $fg = 1_Q$, so f and g are inverse and $P \cong Q$.

17. The pairing of any arrow with the identity is in fact split mono: $\pi_1 \circ \langle 1_A, f \rangle = 1_A$. There is a functor $G : \textbf{Sets} \to \textbf{Rel}$ which is constant on objects and takes $f : A \to B$ to $(\text{im} \langle 1_A, f \rangle) \subseteq A \times B$. It preserves identities since

$G(1_A) \operatorname{im} \langle 1_A, 1_A \rangle = \{\langle a, a \rangle \mid a \in A\} = 1_A \in \mathbf{Rel}$. It preserves composition because for $g : B \to C$, we have

$$G(g \circ f) = \operatorname{im} \langle 1_A, g \circ f \rangle = \{\langle a, g(f(a)) \rangle \mid a \in A\}$$

$$= \{\langle a, c \rangle \mid \exists b \in B.b = f(a) \wedge c = g(b)\}$$

$$= \{\langle b, g(b) \rangle \mid b \in B\} \circ \{\langle a, f(a) \rangle \mid a \in A\}$$

$$= G(g) \circ G(f).$$

Chapter 3

1. In any category $\mathbf{C}$, the diagram

$$A \xleftarrow{\quad c_1 \quad} C \xrightarrow{\quad c_2 \quad} B$$

is a product diagram iff the mapping

$$\hom(Z, C) \longrightarrow \hom(Z, A) \times \hom(Z, B)$$

given by $f \mapsto \langle c_1 \circ f, c_2 \circ f \rangle$ is an isomorphism. Applying this fact to $\mathbf{C}^{\mathrm{op}}$, the claim follows.

2. Say i_{MA}, i_{MB} are the injections into the coproduct $MA + MB$, and η_A, η_B are the injections into the free monoids on A, B. Put $e = [U(i_{MA}) \circ \eta_A, U(i_{MB}) \circ \eta_B]$. Let an object Z and an arrow $f : A + B \to UZ$ be given. Suppose $h : MA + MB \to Z$ has the property that

$$Uh \circ e = f \qquad\qquad (*)$$

Because of the UMP of the coproduct, we have generally that $a \circ [b, c] = [a \circ b, a \circ c]$, and in particular

$$Uh \circ e = [Uh \circ U(i_{MA}) \circ \eta_A, Uh \circ U(i_{MB}) \circ \eta_B]$$

Because this is equal to f, which is an arrow out of $A + B$, and since functors preserve composition, we have

$$U(h \circ i_{MA}) \circ \eta_A = f \circ i_A$$
$$U(h \circ i_{MB}) \circ \eta_B = f \circ i_B$$

where i_A, i_B are the injections into $A + B$. But the UMP of the free monoid implies that $h \circ i_{MA}$ must coincide with the unique $\overline{f \circ i_A}$ that makes the

triangle

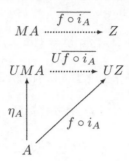

commute. Similarly, $h \circ i_{MB} = \overline{f \circ i_B}$. Since its behavior is known on both injections, h is uniquely determined by the condition $(*)$; in fact, $h = [\overline{f \circ i_A}, \overline{f \circ i_B}]$. That is, the UMP of the free monoid on $A + B$ is satisfied by $MA + MB$. Objects characterized by UMPs are unique up to isomorphism, so $M(A + B) \cong MA + MB$.

5. In the category of proofs, we want to see that (modulo some identifications) the coproduct of formulas φ and ψ is given by $\varphi \vee \psi$. The intro and elim rules automatically give us maps (proofs) of the coproduct from either of its disjuncts, and from pairs of proofs that begin with each of the disjuncts into a single proof beginning with the disjunction. To see that this object really is a coproduct, we must verify that this is the unique commuting arrow.

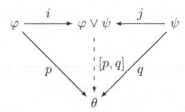

But this is simple since composition is simply concatenation of proofs. Suppose we have another proof $r : \varphi \vee \psi \to \theta$ with $r \circ i = p$. Then by disjunction elimination, r necessarily has the form

$$
\begin{array}{ccc}
& [\varphi] & [\psi] \\
& \vdots & \vdots \\
\varphi \vee \psi & \theta & \theta \\
\hline
& \theta &
\end{array}
$$

Applying i on the right simply has the effect of bringing down part of the proof above, so that the quotienting equation now reads $r \circ i = \begin{matrix}[\varphi] \\ \vdots \\ \theta\end{matrix} = p$. Hence, up to the presence of more detours, we know that the proof appearing as part

$[\psi]$
$\vdots$

of r is exactly p. Similarly, we know that the second part of the proof θ must be q. Thus r is uniquely defined (up to detours) by p and q, $\varphi \vee \psi$ is indeed a coproduct.

6. (Equalizers in **Ab**). Suppose we have a diagram

$$(A, +_A, 0_A) \underset{g}{\overset{f}{\rightrightarrows}} (B, +_B, 0_B)$$

in **Ab**. Put $A' := \{a \in A \mid f(a) = g(a)\}$. It is easy to check that A' is in fact a subgroup of A, so it remains to be shown that

$$(A', +_A, 0_A) \hookrightarrow (A, +_A, 0_A) \underset{g}{\overset{f}{\rightrightarrows}} (B, +_B, 0_B)$$

is an equalizer diagram.

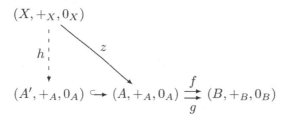

If the triangle is to commute, $h(x) = z(x)$ for all $x \in X$, so h is uniquely determined. It is easily checked that h is a homomorphism, implying that **Ab** indeed has all equalizers.

14. (a) The equalizer of $f \circ \pi_1$ and $f \circ \pi_2$ is the relation $\ker(f) = \{\langle a, a' \rangle \in A \times A \mid f(a) = f(a')\}$. Symmetry, transitivity, and reflexivity of $\ker(f)$ follow immediately from the same properties of equality.

(b) We need to show that a pair a, a' of elements are in the kernel of the projection $q : A \longrightarrow A/R$ iff they are related by R. But this amounts to saying that $q(a) = q(a')$ iff aRa', where $q(x) = \{x \mid xRa\}$ is the equivalence class. But this is true since R is an equivalence relation.

(c) Take any function $f : A \to B$ with $f(a) = f(a')$ for all aRa'. The kernel $\ker(f)$ of f is therefore an equivalence relation that contains R, so $\langle R \rangle \subseteq \ker(f)$. It follows that f factors through the projection $q : A \longrightarrow A/\langle R \rangle$ (necessarily uniquely, since q is epic).

(d) The coequalizer of the projections from R is the projection $q : A \longrightarrow A/\langle R \rangle$, which has $\langle R \rangle$ as its kernel.

Chapter 4

1. Given a categorical congruence $\sim$ on a group G, the corresponding normal subgroup is $N_\sim := \{g \mid g \sim e\}$. N is a subgroup; it contains the identity by reflexivity of $\sim$. It is closed under inverse by symmetry and the fact that $e \sim g$ implies $g^{-1} = g^{-1}e \sim g^{-1}g = e$. It is closed under product because if $g \sim e$ and $h \sim e$ then $gh \sim ge = g \sim e$, and by transitivity $gh \in N_\sim$. It is normal because

$$x\{g \mid g \sim e\} = \{xg \mid g \sim e\} = \{x(x^{-1}h) \mid x^{-1}h \sim e\} = \{h \mid h \sim x\}$$

and

$$\{g \mid g \sim e\}x = \{gx \mid g \sim e\} = \{(hx^{-1})x \mid hx^{-1} \sim e\} = \{h \mid h \sim x\}.$$

In the other direction, the categorical congruence $\sim_N$ corresponding to a normal subgroup N is $g \sim_N h : \iff gh^{-1} \in N$. The fact that $\sim_N$ is an equivalence follows easily from the fact that N is a subgroup. If $f \sim_N g$, then also $hfk \sim_N hgk$, since $fg^{-1} \in N$ implies $hfkk^{-1}g^{-1}h^{-1} = hfg^{-1}h^{-1} \in N$, because N was assumed normal, and so $N = hNh^{-1}$.

Since two elements g, h of a group are in the same coset of N precisely when $gh^{-1} = e$, the quotient G/N and the quotient $G/\sim$ coincide when N and $\sim$ are in the correspondence described above.

6. (a)

$$1 \longrightarrow 2 \longrightarrow 3$$

No equations. (i.e., **3** is free)

(b)

$$1 \underset{g}{\overset{f}{\rightrightarrows}} 2 \longrightarrow 3$$

Equations: $f = g$

(c)

$$1 \underset{g}{\overset{f}{\rightrightarrows}} 2 \underset{k}{\overset{h}{\rightrightarrows}} 3$$

Equations: $f = g$, $h = k$

(d)

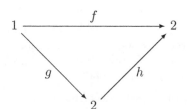

Equations: $f = h \circ g$

7. By definition of congruence, $f \sim f'$ implies $gf \sim gf'$ and $g \sim g'$ implies $gf' \sim g'f'$. By transitivity of $\sim$, we conclude $gf \sim g'f'$.

8. $\sim$ is an equivalence because equality is. For instance, if $f \sim g$, then for all **E** and $H : \mathbf{D} \to \mathbf{E}$ we have $HF = HG \Rightarrow H(f) = H(g)$. But under the same conditions, we have $H(g) = H(f)$, so $g \sim f$. Since H is assumed to be a functor, it preserves composition, and so $H(hfk) = H(h)H(f)H(k) = H(h)H(g)H(k) = H(hgk)$ for any H such that $HF = HG$ and any h, k, hence $\sim$ is a congruence.

   Let q be the functor assigning all the arrows in **D** to their $\sim$-equivalence classes in the quotient $\mathbf{D}/\sim$. We know q is indeed a well-defined functor by a previous exercise. Suppose we have H coequalizing F, G. By definition of $\sim$ any arrows that H identifies are $\sim$-equivalent, and therefore identified also by q. There can be at most one K making the triangle in

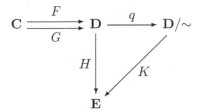

   commute, (for any $[f]_\sim \in \mathbf{D}/\sim$ it must be that $K([f]_\sim) = H(f)$) and the fact that q identifies at least as many arrows as H implies the existence of such a K. So q is indeed the coequalizer of F, G.

Chapter 5

1. Their UMPs coincide. A product in $\mathbf{C}/X$ of f and g is an object $h : A \times_X B \to X$ and projections $\pi_1 : h \to f$ and $\pi_2 : h \to g$ which is terminal among such structures. The pullback of f, g requires an object $A \times_X B$ and projections $\pi_1 : A \times_X B \to A$ and $\pi_2 : A \times_X B \to B$ such that $f \circ \pi_1 = g \circ \pi_2$, terminal among such structures. The commutativity requirements of the pullback are exactly those imposed by the definition of arrow in the slice category.

2. (a) If m is monic, then the diagram is a pullback; if $m \circ f = m \circ g$, then $f = g$, the unique mediating map being equivalently f or g. If the diagram is a pullback, suppose $m \circ f = m \circ g$. The definition of pullback implies the unique existence of h such that $1_M \circ h = f$ and $1_M \circ h = g$, but this implies $f = g$.

3.

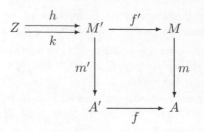

Let $h, k : Z \to M'$ be given. Suppose $m'h = m'k$. Then, $fm'h = fm'k$ and so $mf'h = mf'k$. Since m is assumed mono, $f'h = f'k$. The definition of pullback applied to the pair of arrows $m'k, f'k$ implies, there is exactly one arrow $q : Z \to M'$ such that $m' \circ q = m'k$ and $f' \circ q = f'k$. But both h, k can be substituted for q and satisfy this equation, so $h = k$.

4. Suppose $m : M \to A$ and $n : N \to A$ are subobjects of A. If $M \subseteq N$, then there is an arrow $s : M \to N$ such that $n \circ s = m$. If $z \in_A M$, then there is an arrow $f : Z \to M$ such that $m \circ f = z$. Then $s \circ f$ witnesses $z \in_A N$, since $n \circ s \circ f = m \circ f = z$. If for all $z : Z \to A$ we have $z \in_A M \Rightarrow z \in_A N$, then in particular this holds for $z = m$, and in fact $m \in_A M$ (via setting $f = 1_A$) so $m \in_A N$, in other words $M \subseteq N$.

7. We show that the representable functor $\mathrm{Hom}_{\mathbf{C}}(C, -) : \mathbf{C} \to \mathbf{Sets}$ preserves all small products and equalizers; it follows that it preserves all small limits, since the latter can be constructed from the former. For products, we need to show that for any set I and family $(D_i)_{i \in I}$ of objects of $\mathbf{C}$, there is a (canonical) isomorphism,

$$\mathrm{Hom}(C, \prod_{i \subset I} D_i) \cong \prod_{i \in I} \mathrm{Hom}(C, D_i).$$

But this follows immediately from the definition of the product $\prod_{i \in I} D_i$. For equalizers, consider an equalizer in $\mathbf{C}$,

$$E \xrightarrow{\ e\ } A \underset{g}{\overset{f}{\rightrightarrows}} B.$$

Applying $\mathrm{Hom}(C, -)$ results in the following diagram in $\mathbf{Sets}$:

$$\mathrm{Hom}(C, E) \xrightarrow{\ e_*\ } \mathrm{Hom}(C, A) \underset{g_*}{\overset{f_*}{\rightrightarrows}} \mathrm{Hom}(C, B),$$

which is clearly an equalizer: for, given $h : C \to A$ with $f_*(h) = g_*(h)$, we therefore have $fh = f_*(h) = g_*(h) = gh$, whence there is a unique $u : C \to E$ with $h = eu = e_*(u)$.

8. We have a putative category of partial maps. We need to verify identity and associativity. The first is easy. Any object is a subobject of itself, so we may

set 1_A in the category of partial maps to be the pair $(1_A, A)$. It is trivial to check that this acts as an identity.

For associativity, suppose U, V, and W are subobjects of A, B, and C, respectively, and that we have maps as in the diagram:

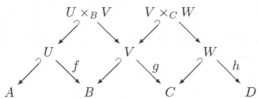

Now let P be the pullback of $U \times_B V$ and $V \times_C W$ over E and k the associated partial map. Since we can compose pullback squares, that means that P is also the pullback of U and $V \times_C W$ over B. Since the latter is the composition of g and h, this means $k = (h \circ g) \circ f$. Similarly, $k = h \circ (g \circ f)$. Hence the composition of partial maps is associative, and this setup does describe a category.

12. If we let the numeral n denote the initial segment of the natural number sequence $\{0 \le 1 \le \ldots \le n\}$, we have a chain of inclusions in **Pos**:

$$0 \to 1 \to 2 \to \ldots \to n \to \ldots.$$

We would like to determine the limit and colimit of the diagram.

For the limit, suppose we have a cone $\zeta_n : Z \to n$. Since 0 is the initial object, ζ_0 is constant, and each map ζ_n has ζ_0 as a factor (this is the cone condition). But each such map simply takes 0 to itself, regarded as an element of n, so that ζ_n is also the constant zero map. So the limit of the diagram can be (anything isomorphic to) the object 0 together with the inclusions $0 \to n$.

Now suppose we have a co-cone $\psi_n : n \to Y$. The co-cone condition implies that ψ_n is simply the restriction of ψ_{m+n} to the subset $n \subseteq n+m$. If $m < n$, then

$$\psi(m) = \psi_m(m) = \psi_n(m) < \psi_n(n)$$

so this is a monotone function. For any other $\varphi : \mathbb{N} \to Y$, there is some n with $\varphi(n) \ne \psi(n) = \psi_n(n)$. Thus, ψ is the unique map factoring the co-cone on Y. Thus, $\omega = \{0 \le 1 \le 2 \le \ldots\}$ together with the evident injections $n \to \omega$ is the colimit of the diagram.

Chapter 6

Notation: If $f : A \to B^C$, then $\mathsf{ev} \circ f \times C = \overline{f} : A \times C \to B$. If $f : A \times C \to B$, then $\lambda f : A \to B^C$.

2. These isomorphism are witnessed by the following pairs: $f : (A \times B)^C \to A^C \times B^C$ defined by $f = \langle \lambda(\pi_1 \circ \overline{1}_{(A \times B)^C}), \lambda(\pi_2 \circ \overline{1}_{(A \times B)^C}) \rangle$ and $f^{-1} : A^C \times B^C \to (A \times B)^C$ defined by $f^{-1} = \lambda \langle \overline{\pi_1}, \overline{\pi_2} \rangle$; and $g : (A^B)^C \to A^{B \times C}$

defined by $g = \lambda(\overline{\mathrm{ev}} \circ \alpha_{(A^B)^C})$ and $g^{-1} : A^{B \times C} \to (A^B)^C$ defined by $g^{-1} = \lambda\lambda(\mathrm{ev} \circ \alpha^{-1}_{A^B \times C})$, where α_Z is the evident isomorphism from associativity and commutativity of the product, up to isomorphism, $Z \times (B \times C) \to (Z \times C) \times B$.

3. The exponential transpose of ev is 1_{B^A}. The exponential transpose of $1_{A \times B}$ is the "partial pairing function" $A \to (A \times B)^B$ defined by $a \mapsto \lambda b : B.\langle a, b \rangle$. The exponential transpose of $\mathrm{ev} \circ \tau$ is the "partial application function" $A \to B^{B^A}$ defined by $a \mapsto \lambda f : B^A.f(a)$.

6. Here we consider the category **Sub**, whose objects are pairs $(A, P \subseteq A)$, and whose arrows $f : (A, P) \to (B, Q)$ are set functions $A \to B$ such that $a \in P$ iff $f(a) \in Q$. This means that an arrow in this category is essentially a pair of arrows $f_1 : P \to Q$ and $f_2 : A \setminus P \to B \setminus Q$; thus, this is (isomorphic to) the category **Sets**/2.

Now, **Sets**/2 is equivalent to the product category **Sets** × **Sets**, by a previous exercise. This latter category is cartesian closed, by the equational definition of CCCs, which clearly holds in the two factors. But equivalence of categories preserves cartesian closure, so **Sub** is also cartesian closed.

10. For products, check that the set of pairs of elements of ωCPOs A and B ordered pointwise, constitutes an ωCPO (with ω-limits computed pointwise) and satisfies the UMP of a product. Similarly, the exponential is the set of continuous monotone functions between A and B ordered pointwise, with limits computed pointwise. In *strict* ωCPOs, by contrast, there is exactly one map $\{\perp\} \to A$, for any object A. Since $\{\perp\} = 1$ is also a terminal object, however, given an exponential B^A there can be only one map $A \to B$, since $\mathrm{Hom}(A, B) \cong \mathrm{Hom}(1 \times A, B) \cong \mathrm{Hom}(1, B^A)$.

11. (a) The identity
$$((p \vee q) \Rightarrow r) \Rightarrow ((p \Rightarrow r) \wedge (q \Rightarrow r))$$
"holds" in any CC poset with joins, that is, this object is equal to the top element 1. Equivalently, from the definition of $\Rightarrow$, we have
$$((p \vee q) \Rightarrow r) \leq ((p \Rightarrow r) \wedge (q \Rightarrow r)),$$
as follows immediately from part (b), which shows the existence of such an arrow in any CCC.

(b) In any category where the constructions make sense, there is an arrow
$$C^{(A+B)} \to C^A \times C^B.$$
Indeed, by the definition of the coproduct, we have arrows $A \to A + B$ and $B \to A + B$, to which we apply the contravariant functor $C^{(-)}$ to obtain maps $C^{(A+B)} \to A^C$ and $C^{(A+B)} \to C^B$. By the UMP of the product, this gives a map $C^{(A+B)} \to C^A \times C^B$, as desired.

13. This can be done directly by comparing UMPs. For a different proof (anticipating the Yoneda lemma), consider, for an arbitrary object X, the bijective

correspondence of arrows,

$$\frac{(A \times C) + (B \times C) \to X}{(A + B) \times C \to X}.$$

This is arrived at via the canonical isos:

$$\mathrm{Hom}((A \times C) + (B \times C), X) \cong \mathrm{Hom}(A \times C, X) \times \mathrm{Hom}(B \times C, X)$$
$$\cong \mathrm{Hom}(A, X^C) \times \mathrm{Hom}(B, X^C)$$
$$\cong \mathrm{Hom}(A + B, X^C)$$
$$\cong \mathrm{Hom}((A + B) \times C, X).$$

Now let $X = (A \times C) + (B \times C)$, respectively $X = (A + B) \times C$, and trace the respective identity arrows through the displayed isomorphisms to arrive at the desired isomorphism

$$(A \times C) + (B \times C) \cong (A + B) \times C.$$

14. If $D = \emptyset$ then $D^D \cong 1$, so there can be no interpretation of $s : D^D \to D$. If $D \cong 1$ then also $D^D \cong 1$, so there are unique interpretations of $s : D^D \to D$ and $t : D \to D^D$. If $|D| \geq 2$ (in cardinality), then $|D^D| \geq |2^D| \geq |\mathcal{P}(D)|$, so there can be no such (split) mono $s : D^D \to D$, by Cantor's theorem on the cardinality of powersets. Thus, the only models can be $D \cong 1$, and in these, clearly all equations hold, since all terms are interpreted as maps into 1.

Chapter 7

1. Take any element $a \in A$ and compute

$$(\mathcal{F}(h) \circ \phi_A)(a) = \mathcal{F}(h)(\phi_A(a))$$
$$= \mathcal{F}(h)(\{\mathcal{U} \in \mathrm{Ult}(A) \mid a \in \mathcal{U}\})$$
$$= \mathcal{P}(\mathrm{Ult}(h))(\{\mathcal{U} \in \mathrm{Ult}(A) \mid a \in \mathcal{U}\})$$
$$= (\mathrm{Ult}(h))^{-1}(\{\mathcal{U} \in \mathrm{Ult}(A) \mid a \in \mathcal{U}\})$$
$$= \{\mathcal{V} \in \mathrm{Ult}(B) \mid a \in \mathrm{Ult}(h)(\mathcal{V})\}$$
$$= \{\mathcal{V} \in \mathrm{Ult}(B) \mid h(a) \in \mathcal{V}\}$$
$$= \phi_B(h(a))$$
$$= (\phi_B \circ h)(a).$$

4. Both functors are faithful. U is full because every monoid homomorphism between groups is a group homomorphism: if $h(ab) = h(a)h(b)$ then $e = h(a^{-1}a) = h(a^{-1})h(a)$ and symmetrically $e = h(a)h(a^{-1})$ and so $h(a^{-1})$ is the inverse of $h(a)$. V is not full; there are set functions between monoids that are not homomorphisms. Only V is surjective on objects (there

are, for example cyclic groups of every cardinality). Only U is injective on objects, since monoid structure uniquely determines inverses, if they exist.

5. It is easy to check that upward-closed sets are closed under unions and finite intersections. The arrow part of the functor A simply takes a monotone function $f : P \to Q$ to itself, construed as a function $f : A(P) \to A(Q)$. Preservation of identities and composition is therefore trivial, but we must check that f is in fact an arrow in **Top**. Let U be an open (that is, upward-closed) subset of $A(Q)$. We must show that $f^{-1}(U)$ is upward-closed. Let $x \in f^{-1}(U)$ and $y \in P$ be given, and suppose $x \leq y$. We know that $f(x) \in U$ and $f(x) \leq f(y)$ since f is monotone. Because U is upward-closed, we have $f(y) \in U$, so $y \in f^{-1}(U)$ and so f is continuous.
   A is trivially faithful. A is also full: Let f be a continuous function $P \to Q$. Put $D := \{q \in Q \mid f(x) \leq q\}$. Since f is continuous and D is upward-closed, $f^{-1}(D)$ is upward-closed. If $x \leq y$ then the fact that $x \in f^{-1}(D)$ implies $y \in f^{-1}(D)$ and so $f(y) \in D$. That is, $f(x) \leq f(y)$. Hence every continuous function $A(P) \to A(Q)$ is a monotone function $P \to Q$.

6. (a) Let the objects of **E** be those of **C**, and identify arrows in **C** if they are identified by F, that is, let **E** be the quotient category of **C** by the congruence induced by F. The functor D is the canonical factorization of F through the quotient.

   (b) Let **E** be the subcategory of **D** whose objects are those in the image of F, and whose arrows are all the **D**-arrows among those objects. Let D be the inclusion of **E** in **D** and E the evident factorization of F through **E**.

   These factorizations agree iff F itself is injective on objects and full.

7. Suppose α is a natural isomoprhism $F \to G : \mathbf{C} \to \mathbf{D}$. Then it has an inverse α^{-1}. Since $\alpha^{-1} \circ \alpha = 1_F$ and $\alpha \circ \alpha^{-1} = 1_G$, it must be that $\alpha_C \circ \alpha_C^{-1} = 1_{GC}$ and $\alpha_C^{-1} \circ \alpha_C = 1_{FC}$. So the components of α are isomorphisms. If conversely all α's components are isomorphisms, then defining $\alpha_C^{-1} = (\alpha_C)^{-1}$ for all $C \in \mathbf{C}$ makes α^{-1} a natural transformation which is α's inverse. For $f : A \to B$, knowing $Gf \circ \alpha_A = \alpha_B \circ Ff$, we compose on the left with α_B^{-1} and on the right with α_A^{-1} to obtain $Ff \circ \alpha_A^{-1} = \alpha_B^{-1} \circ Gf$, the naturality of α^{-1}.
   The same does not hold for monomorphisms. Let **C** be the two-element poset $\{0 \leq 1\}$ and **D** the category

$$A \overset{x}{\underset{y}{\rightrightarrows}} B \overset{f}{\longrightarrow} C$$

such that $fx = fy$. Let F be the functor taking $0 \leq 1$ to $x : A \to B$ and G the functor taking it to $f : B \to C$. There is a natural transformation $\alpha : F \to G$ such that $\alpha_0 = x : A \to B$ and $\alpha_1 = f : B \to C$. The component f of α is not mono, but α itself is; there are no nontrivial natural transformations into F: any $\beta : H \to F$ would have to satisfy a naturality square

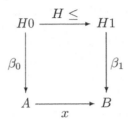

$$
\begin{array}{ccc}
H0 & \xrightarrow{\ H \leq\ } & H1 \\
\downarrow{\beta_0} & & \downarrow{\beta_1} \\
A & \xrightarrow{\ \ x\ \ } & B
\end{array}
$$

But $H0$ must be A and $\beta_0 = 1_A$. Then $H1$ must be either A or B, forcing β to either be the unique natural transformation to F from the functor taking $0 \leq 1$ to $1_A : A \to A$, or else the identity natural transformation on F.

8. Put $(F \times G)(C) = FC \times GC$, and $(F \times G)(f) = Ff \times Gf$. Define $(\pi_1)_C = \pi_1^{FC \times GC} : FC \times GC \to FC$ and $(\pi_2)_C = \pi_2^{FC \times GC} : FC \times GC \to GC$. It is easy to check that π_1, and π_2 are natural. Let a functor $Z : \mathbf{C} \to \mathbf{D}$ and natural transformations $\alpha : Z \to F$ and $\beta \to F$ be given. By the UMP of the product, there are unique arrows $h_C : ZC \to FC \times GC$ such that $(\pi_1)_C \circ h_C = \alpha_C$ and $(\pi_2)_C \circ h_C = \beta_C$. We need to verify that

$$
\begin{array}{ccc}
ZC & \xrightarrow{\ h_C\ } & FC \times GC \\
\downarrow{Zf} & & \downarrow{Ff \times Gf} \\
ZD & \xrightarrow[\ h_D\ ]{} & FD \times GD
\end{array}
$$

But

$$
\pi_1^{FD \times GD} \circ Ff \times Gf \circ h_C = Ff \circ \pi_1^{FC \times GC} \circ h_C
$$
$$
= Ff \circ \alpha_C = \alpha_D \circ Zf = \pi_1^{FD \times GD} \circ h_D \circ Zf.
$$

And similarly with the second projection, using the naturality of β.

10. To satisfy the bifunctor lemma, we need to show that for any $f : C \to C' \in \mathbf{C}^{\mathrm{op}}$ and $g : D \to D' \in \mathbf{C}$ the following commutes:

$$
\begin{array}{ccc}
\hom(C, D) & \xrightarrow{\ \hom(f, D)\ } & \hom(C', D) \\
\downarrow{\hom(C, g)} & & \downarrow{\hom(C', g)} \\
\hom(C, D') & \xrightarrow[\ \hom(f, D')\ ]{} & \hom(C', D')
\end{array}
$$

But either path around the square takes an arrow $h : C \to D$ and turns it into $g \circ h \circ f : C' \to D'$; thus the associativity of composition implies that the square commutes.

12. If $\mathbf{C} \simeq \mathbf{D}$, then there are functors $F : \mathbf{C} \rightleftarrows \mathbf{D} : G$ and natural isomorphisms $\alpha : 1_{\mathbf{D}} \to FG$ and $\beta : GF \to 1_{\mathbf{C}}$. Suppose $\mathbf{C}$ has products, and let $D, D' \in \mathbf{D}$ be given. We claim that $F(GD \times GD')$ is a product object of D and D', with projections $\alpha_D^{-1} \circ F\pi_1^{GD \times GD'}$ and $\alpha_{D'}^{-1} \circ F\pi_2^{GD \times GD'}$. For suppose we have an object Z and arrows $a : Z \to D$ and $a' : Z \to D'$ in $\mathbf{D}$. There is a unique $h : GZ \to GD \times GD' \in \mathbf{C}$ such that $\pi_1^{GD \times GD'} \circ h = Ga$ and $\pi_2^{GD \times GD'} \circ h = Ga'$. Then the mediating map in $\mathbf{D}$ is $Fh \circ \alpha_Z$. We can calculate

$$
\begin{aligned}
\alpha_D^{-1} \circ F\pi_1^{GD \times GD'} \circ Fh \circ \alpha_Z &= \alpha_D^{-1} \circ F(\pi_1^{GD \times GD'} h) \circ \alpha_Z \\
&= \alpha_D^{-1} \circ FGa \circ \alpha_Z \\
&= \alpha_D^{-1} \circ \alpha_D \circ a \\
&= a
\end{aligned}
$$

and similarly for the second projection.

Uniqueness of the map $Fh \circ \alpha_Z$ follows from that of h.

16. Let $\mathbf{C}$ be given. Choose one object $D_{[C]_\cong}$ from each isomorphism class $[C]_\cong$ of objects in $\mathbf{C}$ and call the resulting full subcategory $\mathbf{D}$. For every object C of $\mathbf{C}$ choose an isomorphism $i_C : C \to D_{[C]_\cong}$. Then, $\mathbf{C}$ is equivalent to $\mathbf{D}$ via the inclusion functor $I : \mathbf{D} \to \mathbf{C}$ and the functor F defined by $FC = D_{[C]_\cong}$ and $F(f : A \to B) = i_B \circ f \circ i_A^{-1}$ (F is a functor because the i_Cs are isomorphisms) and i construed as a natural isomorphism $1_{\mathbf{D}} \to FI$ and $1_{\mathbf{C}} \to IF$. Naturality is easy to check:

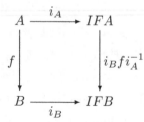

So $\mathbf{C}$ is equivalent to the skeletal category $\mathbf{D}$.

Chapter 8

1. Let $f : C \rightleftarrows C' : g$ be an iso. Then, clearly, $Ff : FC \rightleftarrows FC' : Fg$ is also one. Conversely, if $p : FC \rightleftarrows FC' : q$ is an iso, then since F is full there are $f : C \rightleftarrows C' : g$ with $Ff = p$ and $Fg = q$. Then $g \circ f = 1_C$ since $F(g \circ f) = Fg \circ Ff = 1_{FC} = F(1_C)$, and F is faithful. Similarly, $f \circ g = 1_{C'}$.

2. Given two natural transformations $\varphi, \psi : P \to Q$, where $P, Q \in \mathbf{Sets}^{\mathbf{C}^{\mathrm{op}}}$, assume that for each $C \in \mathbf{C}$ and $\theta : yC \to P$, we have $\varphi \circ \theta = \psi \circ \theta$. In other words,

$$\varphi_C * = \psi_C * : \hom(yC, P) \to \hom(yC, Q).$$

The Yoneda lemma gives us a bijection $\hom(yC, P) \cong PC$ for each C, and these bijections are natural in P, so the following diagram commutes:

$$
\begin{array}{ccc}
\hom(yC, P) & \xrightarrow{\;\cong\;} & PC \\[4pt]
\Big\downarrow{\scriptstyle \varphi_C* \,=\, \psi_C*} & \quad {\scriptstyle \varphi_C}\Big\downarrow\Big\downarrow{\scriptstyle \psi_C} & \\[4pt]
\hom(yC, Q) & \xrightarrow[\;\cong\;]{} & QC
\end{array}
$$

But then both φ_C and ψ_C must be given by the single composition through the left side of the square, so that $\varphi = \psi$.

3. The following isos are natural in Z:

$$
\begin{aligned}
\hom_{\mathbf{C}}(Z, A^B \times A^C) &\cong \hom_{\mathbf{C}}(Z, A^B) \times \hom_{\mathbf{C}}(Z, A^C) \\
&\cong \hom_{\mathbf{C}}(Z \times B, A) \times \hom_{\mathbf{C}}(Z \times C, A) \\
&\cong \hom_{\mathbf{C}}((Z \times B) + (Z \times C), A) \\
&\cong \hom_{\mathbf{C}}(Z \times (B + C), A) \\
&\cong \hom_{\mathbf{C}}(Z, A^{B+C}).
\end{aligned}
$$

Hence $A^B \times A^C \cong A^{B+C}$, since the Yoneda embedding is full and faithful. The case of $(A \times B)^C \cong A^C \times B^C$ is similar.

6. Limits in functor categories $\mathbf{D}^{\mathbf{C}}$ can be computed "pointwise": given $F : \mathbf{J} \to \mathbf{D}^{\mathbf{C}}$ set

$$(\varprojlim_{j \in \mathbf{J}} Fj)(C) = \varprojlim_{j \in \mathbf{J}}(Fj(C)).$$

Thus, it suffices to have limits in $\mathbf{D}$ in order to have limits in $\mathbf{D}^{\mathbf{C}}$. Colimits in $\mathbf{D}^{\mathbf{C}}$ are limits in $(\mathbf{D}^{\mathbf{C}})^{\mathrm{op}} = (\mathbf{D}^{\mathrm{op}})^{\mathbf{C}^{\mathrm{op}}}$.

7. The following are natural in C:

$$
\begin{aligned}
y(A \times B)(C) &\cong \hom(C, A \times B) \\
&\cong \hom(C, A) \times \hom(C, B) \\
&\cong y(A)(C) \times y(B)(C) \\
&\cong (y(A) \times y(B))(C),
\end{aligned}
$$

so $y(A \times B) \cong y(A) \times y(B)$. For exponentials, take any A, B, C and compute:

$$y(B)^{y(A)}(C) \cong \hom(yC, yB^{yA})$$
$$\cong \hom(yC \times yA, yB)$$
$$\cong \hom(y(C \times A), yB)$$
$$\cong \hom(C \times A, B)$$
$$\cong \hom(C, B^A)$$
$$\cong y(B^A)(C).$$

12. (a) For any poset $\mathbf{P}$, the subobject classifier Ω in $\mathbf{Sets^P}$ is the functor:

$$\Omega(p) = \{F \subseteq \mathbf{P} \mid (x \in F \Rightarrow p \leq x) \wedge (x \in F \wedge x \leq y \Rightarrow y \in F)\},$$

that is, $\Omega(p)$ is the set of all upper sets above p. The action of Ω on $p \leq q$ is by "restriction": $F \mapsto F|_q = \{x \in F \mid q \leq x\}$. The point $t : 1 \to \Omega$ is given by selecting the maximal upper set above p,

$$t_p(*) = \{x \mid p \leq x\}.$$

In $\mathbf{Sets^2}$, the subobject classifier is therefore the functor $\Omega : \mathbf{2} \to \mathbf{Sets}$ defined by

$$\Omega(0) = \{\{0, 1\}, \{1\}\}$$
$$\Omega(1) = \{\{1\}\},$$

together with the natural transformation $t : 1 \to \Omega$ with

$$t_0(*) = \{0, 1\}$$
$$t_1(*) = \{1\}.$$

In $\mathbf{Sets}^\omega$, the subobject classifier is the functor $\Omega : \omega \to \mathbf{Sets}$ defined by

$$\Omega(0) = \{\{0, 1, 2, \dots\}, \{1, 2, 3, \dots\}, \{2, 3, 4, \dots\}, \dots\}$$
$$\Omega(1) = \{\{1, 2, 3, \dots\}, \{2, 3, 4, \dots\}, \{3, 4, 5, \dots\}, \dots\}$$
$$\vdots = \vdots$$
$$\Omega(n) = \{\{n, n+1, n+2, \dots\}, \{n+1, n+2, n+3, \dots\}, \dots\},$$

with the transition maps $\Omega(n) \to \Omega(n+1)$ defined by taking $\{n, n+1, n+2, \dots\}$ to $\{n+1, n+2, n+3, \dots\}$ and like sets to themselves, together with the natural transformation $t : 1 \to \Omega$ with

$$t_0(*) = \{0, 1, 2, \dots\}$$
$$t_1(*) = \{1, 2, 3, \dots\}$$
$$t_n(*) = \{n, n+1, n+2, \dots\}.$$

(b) One can check directly that all of the topos operations—pullbacks, exponentials, subobject classifier—construct only finite set-valued functors
when applied to finite set-valued functors.

Chapter 9

3. η_A takes an element $a \in A$ and returns the function $(c \mapsto \langle a, c \rangle) \in (A \times C)^C$.

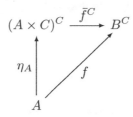

4. For any small index category $\mathbf{J}$, the left adjoint of $\Delta : \mathbf{C} \to \mathbf{C}^{\mathbf{J}}$ is the functor
taking a diagram in $\mathbf{C}^{\mathbf{J}}$ to its colimit (if it exists), and the right adjoint to
its limit. Indeed, suppose $D : \mathbf{J} \to \mathbf{C}$ is a functor.

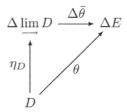

Define the natural transformation η_D to take an object $J \in \mathbf{J}$ to the injection
$i_J : DJ \to \varinjlim D$. The commutativity condition on the colimit guarantees
that η_D is natural. Suppose E and $\theta : D \to \Delta E$ are given. That is, suppose θ
is a co-cone from the diagram D to the object E. Then there exists a unique
arrow out of $\bar{\theta} : \varinjlim D \to E$ making the above diagram commute. Therefore,
$\varinjlim \dashv \Delta$. Dually, $\Delta \dashv \varprojlim$.
It follows that for $\mathbf{J} = \mathbf{2}$, the left adjoint is binary coproduct and the right
adjoint is binary product.

5. Right adjoints preserve limits, and left adjoints preserve colimits.

8. The first adjunction is equivalent to the statement:

$$\mathrm{im}(f)(X) \subseteq Y \iff X \subseteq f^{-1}(Y),$$

for all $X \subseteq A, Y \subseteq B$. Here,

$$\mathrm{im}(f)(X) = \{b \mid b = f(x) \text{ for some } x \in X\}$$

$$f^{-1}(Y) = \{a \mid f(a) \in Y\}$$

If $\mathrm{im}(f)(X) \subseteq Y$ then for any $x \in X$, we have $f(x) \in Y$, and so $X \subseteq f^{-1}(Y)$.
Conversely, take $b \in \mathrm{im}(f)(X)$, so there is some $x \in X$ with $f(x) = b$. If
$X \subseteq f^{-1}(Y)$ then $b = f(x) \in Y$.

For the right adjoint, set

$$f_*(X) = \{b \mid f^{-1}(\{b\}) \subseteq X\}.$$

We need to show

$$f^{-1}(Y) \subseteq X \iff Y \subseteq f_*(X).$$

Suppose $f^{-1}(Y) \subseteq X$ and take any $y \in Y$, then $f^{-1}(\{y\}) \subseteq f^{-1}(Y) \subseteq X$. Conversely, given $Y \subseteq f_*(X)$, we have $f^{-1}(Y) \subseteq f^{-1}(f_*(X)) \subseteq X$, since $b \in f_*(X)$ implies $f^{-1}(\{b\}) \subseteq X$.

9. We show that $\mathcal{P} : \mathbf{Sets}^{\mathrm{op}} \to \mathbf{Sets}$ has itself, regarded as a functor $\mathcal{P}^{\mathrm{op}} : \mathbf{Sets} \to \mathbf{Sets}^{\mathrm{op}}$, as a (left) adjoint:

$$\mathrm{Hom}_{\mathbf{Sets}}(A, \mathcal{P}(B)) \cong \mathrm{Hom}_{\mathbf{Sets}}(A, 2^B) \cong \mathrm{Hom}_{\mathbf{Sets}}(B, 2^A)$$

$$\cong \mathrm{Hom}_{\mathbf{Sets}}(B, \mathcal{P}(A)) \cong \mathrm{Hom}_{\mathbf{Sets}^{\mathrm{op}}}(\mathcal{P}^{\mathrm{op}}(A), B).$$

10. A right adjoint to $U : \mathbf{C}/C \to \mathbf{C}$ is given by products with C,

$$A \mapsto (\pi_2 : A \times C \to C),$$

so U has a right adjoint iff every object A has such a product.

To have a left adjoint, U would have to preserve limits, and in particular the terminal object $1_C : C \to C$. But $U(1_C) = C$, so C would need to be terminal, in which case $\mathbf{C}/C \cong \mathbf{C}$.

11. (a) In a Heyting algebra, we have an operation $b \Rightarrow c$ such that

$$a \leq b \Rightarrow c \iff a \wedge b \leq c.$$

We define a coHeyting algebra by duality, as a bounded lattice with an operation a/b satisfying

$$a/b \leq c \iff a \leq b \vee c$$

In a Boolean algebra, we know that $b \Rightarrow c = \neg b \vee c$. By duality, we can set $a/b = a \vee \neg b$.

(b) In intuitionistic logic, we have two inference rules regarding negation:

$$\varphi \wedge \neg\varphi \vdash \bot$$

$$\varphi \vdash \neg\neg\varphi$$

We get inference rules for the conegation $\sim p = 1/p$ by duality

$$\top \vdash \varphi \vee \sim \varphi$$

$$\sim\sim \varphi \vdash \varphi$$

For the boundary $\partial p = p \wedge \sim p$, we have the inference rules derived from the rules for $\wedge$:

$$q \vdash \partial p \quad \text{iff} \quad q \vdash p \text{ and } q \vdash \sim p$$

(c) We seek a biHeyting algebra P which is not Boolean. The underlying lattice of P will be the three-element set $\{0, p, 1\}$, ordered $0 \leq p \leq 1$. Now let

$$x \Rightarrow y = \begin{cases} 1 & x \leq y \\ y & \text{o.w.} \end{cases}$$

This is easily checked to satisfy the required condition for $x \Rightarrow y$, thus P is a Heyting algebra. But since P is self-dual, it is also a coHeyting algebra, and co-implication must be given by

$$x/y = \begin{cases} 0 & x \geq y \\ y & \text{o.w.} \end{cases}$$

To see that P is not Boolean, observe that $\neg x = x \Rightarrow 0 = 0$, so $\neg\neg x = 1 \neq x$.

Note that P is the lattice of lower sets in the poset $\mathbf{2}$. In general, such a lattice is always a Heyting algebra, since it is completely distributive, as is easily seen. It follows that such a lattice is also coHeyting, since its opposite is isomorphic to the lower sets in the opposite of the poset.

19. The right adjoint $\mathbf{Rel} \to \mathbf{Sets}$ is the powerset functor, $A \mapsto \mathcal{P}(A)$, with action on a relation $R \subseteq A \times B$ given by

$$\mathcal{P}(R) : \mathcal{P}(A) \to \mathcal{P}(B)$$

$$X \mapsto \{b \mid xRb \text{ for some } x \in X\}.$$

The unit $\eta_A : A \to \mathcal{P}(A)$ is the singleton mapping $a \mapsto \{a\}$, and the counit is the (converse) membership relation $\ni_A \subseteq \mathcal{P}(A) \times A$.

Chapter 10

2. Let $\mathbf{C}$ be a category with terminal object 1 and binary coproducts, and define $T : \mathbf{C} \to \mathbf{C}$ by $TC = 1 + C$. Let $\mathbb{T}$ be the equational theory of a set equipped with a unary operation and a distinguished constant (no equations). We want to show that the following categories are equivalent:

T-algebras	Objects :	$(A \in \mathbf{C}, a : 1 + A \to A)$
	Arrows :	$h : (A, a) \to (B, b)$ s.t. $h \circ a = b \circ T(h)$
$\mathbb{T}$-algebras	Objects :	$(X \in \mathbf{Sets}, c_X \in X, s_X : X \to X)$
	Arrows :	$f : X \to Y$ s.t $f c_X = c_Y$ and $f \circ s_X = s_Y \circ f$

We have the functor $F : T\text{-Alg} \to \mathbb{T}\text{-Alg}$ sending

$$(A, a) \mapsto (A, a_1 : 1 \to A, a_2 : A \to A)$$

where $a = [a_1, a_2]$ as a map from the coproduct $1 + A$.

Conversely, given $(X, c \in X, s : X \to X)$, we can set $f = [c, s] : 1 + X \to X$ to get a T-algebra. The effect on morphisms is easily seen, as is the fact that these are pseudo-inverse functors.

Since free $\mathbb{T}$-algebras exist in **Sets** and such existence is preserved by equivalence functors, it follows that **Sets** has free T-algebras. In particular, an initial T-algebra in **Sets** is the initial $\mathbb{T}$-algebra $\mathbb{N}$, which is an NNO.

3. Let $i : TI \to I$ be an initial T-algebra. By initiality, we can (uniquely) fill in the dotted arrows of the following diagram:

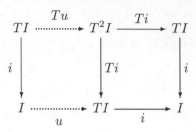

Composing the squares, we have a map of T-algebras $I \to I$, which by uniqueness must be the identity. But then $i \circ u = 1_I$, and $u \circ i = Ti \circ Tu = T(i \circ u) = 1_{TI}$, so i is an isomorphism. A natural numbers object N is initial for the endofunctor $TC = 1 + C$, so it follows that $N \cong 1 + N$ for any NNO.

REFERENCES

The following works are referred to in the text. They are also recommended for
further reading.

1. Eilenberg, S. and S. Mac Lane (1945) "General theory of natural equivalences," *Transactions of the American Mathematical Society* 58, 231–94.

2. Johnstone, P.T. (1982) *Stone Spaces*, Cambridge: Cambridge University
   Press.

3. Johnstone, P.T. (2002) *Sketches of an Elephant: A Topos Theory Compen-
   dium*, 2 vols, Oxford: Oxford University Press.

4. Lambek, J. and P. Scott (1986) *Introduction to Higher-Order Categorical
   Logic*, Cambridge: Cambridge University Press.

5. Lawvere, F.W. (1969) "Adjointness in Foundations," *Dialectica*, 23, 281–96.

6. Mac Lane, S. (1971) *Categories for the Working Mathematician*, Springer:
   Berlin, Heidelberg, New York, 2nd ed. 1998.

7. Mac Lane, S. and I. Moerdijk (1992) *Sheaves in Geometry and Logic: A First
   Introduction to Topos Theory*, Springer: Berlin, Heidelberg, New York.

8. McLarty, C. (1995) *Elementary Categories, Elementary Toposes*, Oxford:
   Oxford University Press.

INDEX